POPULATION ECOLOGY OF HUMAN SURVIVAL

The Location of the 13 Gidra Villages

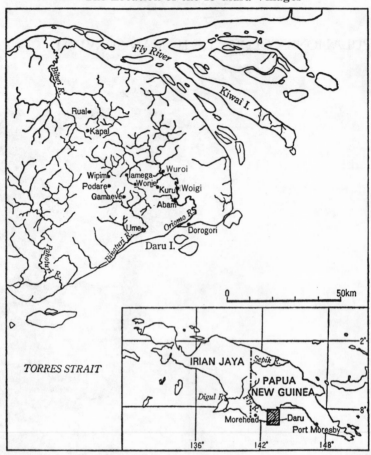

Age-Grade System of the Gridra Society

Male	Estimated Age	Female
Miid	50s	Nanyukonga
Nanyuruga		
	40s	
Rugajog		Kongajog
	20s	
Kewalbuga		
	16–17	
Yambuga		Ngamugaibuga
	7–8	
Sobijogbuga		Sobijogngamugai

POPULATION ECOLOGY OF HUMAN SURVIVAL

Bioecological Studies of the Gidra in Papua New Guinea

Edited by Ryutaro Ohtsuka and Tsuguyoshi Suzuki

UNIVERSITY OF TOKYO PRESS

Publication of this book was supported by a grant-in-aid from the Nippon Life Insurance Foundation.

© UNIVERSITY OF TOKYO PRESS, 1990
ISBN 4-13-066108-6
ISBN 0-86008-456-6

Printed in Japan.

CONTENTS

CONTENTS

PREFACE

Field scientists who study adaptation and survival of human beings owe a great deal to the people among whom they conduct their research. We learned a number of things, both factual and bearing on the research framework, from the Gidra-speaking Papuans, the target group of our study and thus of this book. Among those, the most stimulating was discovering the important role that space and time dimensions play in the Gidra people's adaptation to their environment. If a single researcher conducts a survey (of one year, for example) in a single village as a participant observer, his findings can tell how the subject villagers manage to live during that period, but not how the neighboring groups who have the same culture but different environments do, nor how the groups as a whole are interrelated. Furthermore, when the fact that the living conditions he observes have resulted from a target people's cumulative history of living is appreciated, collection of long-term data (of covering the period over years or human generations) becomes necessary. These points were brought home to us in reality as well as theory when one of us (R.O.) completed research in one of the Gidra villages in 1971-72 and we discussed his findings.

As human ecologists, all the contributors to this book have long been interested in population ecology, which traditionally progressed through the study of animal ecology rather than human ecology. All of us understood that the insights obtained from R.O.'s research in 1971-72 overlapped with the concerns of population ecology. Consequently, we organized a new project, and five of us (T.S., R.O., T.A., T.K., and T.I.) conducted field work in 1980 and 1981-82 among the Gidra people as a whole. The findings of this second research project are the major data source for this volume; in addition, it includes some of the data from R.O.'s survey in 1971-72 and some from a third and ongoing project, in which all of us are now engaged.

The Gidra people are ideally suited to our focus on human population ecology in two particular respects. One is that they have survived in small numbers, fewer than 2000, fairly independently of adjacent groups in their means of subsistence (production and consumption) as well as their marriage organization; this population size made it possible for us to study the whole pupulation. The second aspect is that their villages, which can be regarded as sub-populations, are located in various environments, and this situation gave us a chance to look at the infrastructure of a human population. It should also be noted, however, that some characteristics of this group are not suited to our investigations. The lack of quantitative information in the past and the difficulty of determining the people's ages are examples. Thus, in some cases we were

vii

obliged to find ways which could be substituted for the standard ones, if less than satisfactorily.

The research methodologies that would be applied to our project were an important concern. Field scientists have developed a variety of research methods, including those which originated in other scientific fields. Interviewing subjects has long been a principal tool of sociocultural scientists. Observation of behavior has also been popular in sociocultural and ecological studies. Human biologists depend on measurements of body physique, physiological capabilities, and genetic features. At the same time, these scholars collect samples of subjects' blood, urine, and hair, the foods they eat, and the water and soil in their environment, and analyze them in the laboratory. Each of these methods has its own specific purposes and priorities. In the field of human ecology, which aims at a "holistic" approach to human adaptation and survival, as we believe, it is natural to use a number of methodologies simultaneously in the research. We attempted to apply as many appropriate methods as we could, unless they were rejected by the people themselves or were inapplicable due to local conditions like unavailability of electricity or roads for vehicles.

The contents of most of the chapters in this book have already appeared elsewhere, and we are very grateful to the publishers and scientific associations concerned for permission to reproduce these materials; they are listed on pages 259-260. Our studies were performed with the collaboration of members of our research project or contributors to this book. However, each chapter in this volume was revised and rearranged by one or two of us.

Our sincere thanks go to the following funding bodies, whose financial support made possible our field work among the Gidra Papuans. The 1971-72 survey was supported by the Wenner-Gren Foundation for Anthropological Research, under the project title "Functional-Ecological Study of Material Conditions of Man's Life in Tropical Habitat, Papua" (project leader: Hitoshi Watanabe). The 1980 and 1981-82 surveys, under the title "Human Adaptability in Lowlands of the South Pacific" (project leader: Tsuguyoshi Suzuki), were done with financial support from Japan's Ministry of Education, Science and Culture, the Monbusho. Our third and ongoing project, "Comparative Ecology of Adaptive Mechanisms of Human Populations in a Diversified Melanesian Environment" (project leader: Ryutaro Ohtsuka), which began in 1985 and aims to compare the adaptive mechanisms among the Gidra and several other Papua New Guinean populations, has also been supported by the Monbusho. Publication of this book was financially supported by the Nippon Life Insurance Foundation, to which we are also very grateful.

The actual field surveys were made possible by the suggestions and support given us by many scholars and Papua New Guinea government officials. Hitoshi Watanabe initiated the field study in the Oriomo Plateau, Papua New Guinea. For this survey in 1971-72 and the preceding short reconnaissance survey in

1967, R.O. owes thanks, in particular, to Jack Golson, D.J. Mulvaney, R.G. Crocombe, Robin Hide, Marion W. Ward, David A.M. Lea, Ralph Bulmer, R.Gerald Ward, and Fred Parker. For the surveys since 1980, we are especially indebted to Andrew Strathern, D.G. Townsend, Richard Jackson, and Jacob L. Simet. Botanical specimens were identified by the staff of the Office of Forests in Lae, Papua New Guinea. The composition of the food samples was partly analyzed by Tatsuyuki Sugahara, and the elemental analyses of the samples of scalp hair and drinking water were conducted with the assistance of Masatoshi Morita.

Finally but mostly, our hearty thanks should go to all of the Gidra-speaking people, who generously accepted us and participated in our various investigations. It is impossible for us to mention so many names of the persons who afforded thoughtful assistances, and we only wish that they could know our appreciation for their friendship and tolerance.

October 1989   *Ryutaro Ohtsuka*
                                              Tsuguyoshi Suzuki

LIST OF CONTRIBUTORS

Tomoya Akimichi is Associate Professor at the National Museum of Ethnology in Osaka. He has been engaged in field investigations in several regions throughout Oceania, and his major concern is biocultural and behavioral adaptation. In the Gidra project, most of his time was spent in Ume village.

Tetsuro Hongo is Instructor in the Department of Human Ecology, School of Health Sciences, Faculty of Medicine, University of Tokyo. Through laboratory and field works, he aims to analyze element and/or micronutrient dynamics in the man-environment relationship.

Tsukasa Inaoka is Instructor in the Department of Public Health, Kumamoto University School of Medicine. He has majored physiological and chemical analysis in relation to nutritional adaptation. In the Gidra project, he took initiative for the study of Dorogori villagers.

Toshio Kawabe is Instructor in the Department of Human Ecology, School of Health Sciences, Faculty of Medicine, University of Tokyo. His research focus is morphological adaptation and human growth study, and in the Gidra project he stayed for long period in Rual village.

Ryutaro Ohtsuka is Associate Professor in the Department of Human Ecology, School of Health Sciences, Faculty of Medicine, University of Tokyo. He is concerned in integrated approach to human adaptive mechanisms in regional ecosystems, paying special attention to subsistence, nutrition, and demography. Since 1971, he has investigated the Gidra people, villagers in Wonie in particular.

Tsuguyoshi Suzuki is Professor in the Department of Human Ecology, School of Health Sciences, Faculty of Medicine, University of Tokyo. He has worked in various fields of health sciences (in particular, occupational health and environmental health) and has been devoted to human ecology in recent years. In the Gidra project, his interest centered on ecology of nutrition and health.

POPULATION ECOLOGY OF HUMAN SURVIVAL

Toward Human Population Ecology

The principal goal of human ecology, or the ecological approach to mankind, is to understand human adaptive mechanisms in relation to the environmental setting. Major efforts have been devoted to field research to identify a variety of man-environment relationships in regional ecosystems. Along this line, the man-environment relationship has been defined, with some concrete aspects of human activities as the linkage: for instance, time allocation (e.g. Erasmus, 1955; Carneiro, 1957; Johnson, 1975; Watanabe 1977; Gross, 1984; Moji, 1987) and energy balance (e.g. Odum, 1967; Rappaport, 1968; Thomas, 1973; Smith, 1979; Hames and Vickers, 1983; Kuchikura, 1988). Most of these studies focused on quantitative grasp of human adaptive mechanisms through direct observation and measurement.

However, these studies looked at human adaptation only for the limited time during which the field investigation was undertaken. According to Ellen (1982), the duration in which human adaptation is assessed can be classified into three categories: short-term within an annual cycle, medium-term within a single life span, and long-term over generations. The medium-term and long-term adaptation has more validity, and some of the researchers made efforts to obtain data relevant to this purpose, with various degrees of reliability. Such efforts met with success to different degrees, depending largely on the conditions of the target human group. In our research project among the Gidra-speaking people who have survived in nonliterate conditions, quantitative data have been favored; data for medium- and long-term adaptation, though only for several aspects, were collected by means of repetition of investigation on the one hand and on the other hand by estimating the past conditions from reliable information collected in the field.

Bearing this notion in mind, it should also be noted that the measures for evaluating human adaptation differ according to which organizational level is dealt with. In our view (e.g. Suzuki, 1977; Ohtsuka, 1987; Ohtsuka and Suzuki, 1989), human adaptation can be assessed at the three following levels, as in animal ecology (e.g. Odum, 1971): individual, population, and ecosystem levels. The approach to man's adaptation at

3

the ecosystem level is effective to elucidate the effects of human activities on the ecosystem, for instance, on the energy flow and material (elemental) cycling, and thus is essential to assess human survival in the global system, particularly in view of environmental deterioration, population increases, and problems in food productivity in the long term or in the future. However, such approaches tended to overlook the interwoven network of biological, sociocultural, technological, and behavioral characteristics of the people and environmental characteristics in local ecosystems, to which special attention was paid in the field investigations of human ecology. In our research project, energy and nutrients were studied in relation to nutritional and subsistence adaptation at a population or an individual level rather than at an ecosystem level.

Man's adaptation at a population level is difficult to distinguish from that at an individual level. By using the term population ecology, we intend to clarify how humans adapt to their environment, taking the time and space dimensions into account. The term population sometimes refers to the number of people itself, as in demography. In the field of biology, the same term denotes a biotic population. In short, a biotic population (of animal species) is defined genetically as "a reproductive community of sexual and cross-fertilizing individuals which share a common gene pool" (Dobzhansky, 1968), and ecologically as "a collective group of organisms of the same species occupying a particular space" (Odum, 1971). It is also recognized that members of a population are small in number and the space is narrow in area. These two definitions look at different domains of an animal population, but they may coexist in many cases. These definitions are appropriate to our species, and applicable at least theoretically and actually in some cases, although the structure and function markedly vary from one human population to another, compared to the other animals.

One may still question whether the geneticists' definition of population is applicable to a small human group (several hundred or a few thousand in number), if it occupies a narrow bounded space, as in the ecologists' definition of population. In fact, however, endogamous rates observed in some of the so-called anthropological populations are so high as to be treated as genetically defined populations. The Gidra, our target group, form one such population, as explained later in detail.

Some of the genetics-oriented studies, based on empirical data, have paid special attention to the population structure (e.g. Dyke, 1971; Leslie, 1985). For instance, Dyke (1984) has proposed an analytical model in which the population structure comprises many interacted subdivisions such as genetic, demographic (in terms of fertility and mortality), mating, kinship, social, and geographic structures. One of the most stimulating points in our research framework is that the genetic structure is closely associated with other structures, particularly the mating structure. In a broader sense, this idea emphasizes the central role of population reproduction or inter-generational replacement of members in relation to marriage organization in the human population structure.

In comparison with the term "society," Ellen (1982: 77) pointed out the preferability of "ecologically defined population" in understanding human adaptation and evolution, since the population is "a more-or-less bounded unit, subject to some degree of quantitative description and analysis." Agreeing with his point, we have also been intrigued by the findings of animal population ecologists that physiological features are diversified either among individuals or among sub-groups in a population, and such diversities are beneficial for the survival of a population as a whole in coping with environmental diversity and fluctuations. In the case of human populations, the environmental (natural and man-made) conditions within the "bounded" space are also more or less heterogeneous and fluctuating. Thus, the diversity in adaptive mechanisms among sub-groups should be carefully examined. In other words, one of our intentions in the population ecology approach is to assess the parameters of adaptation not by the mean values of the members as a whole or broken down by such categories as sex and age, but by the values differing among the sub-groups in relation to the environmental diversity.

Another characteristic of our research project is that it pays special attention to health and nutrition. Human nutrition, even in a narrower sense, is tightly associated with the way by which the people adapt to their habitat through exploiting food resources (Weiner, 1977), and consequently human nutritional conditions can be recognized as good measures of adaptation. Looking from another angle, nutritional status is one of the determinants of human physique, physiological capability, activity pattern, and reproduction (Suzuki, 1984). Similarly, human health status is considered a measure of the effectiveness with which human groups adapt to their environment and, concurrently, respond to their environment (Lieban, 1973; Montogomery, 1973; Wellin, 1978). Furthermore, Suzuki (1987) has developed these ecology-oriented concepts, and suggests that human health at a population level can be evaluated by the measures which are also applicable to the evaluation of population survival; three interrelated aspects, i.e. production, consumption, and reproduction, are basic for these evaluations, with close relation to the human ecological complex consisting of natural conditions, technology, social organization, and cognitive structure of the environment.

Finally, a brief explanation should be given of why the Gidra-speaking people were suitable as our research subjects. As was suggested by Bayliss-

Smith (1977), Melanesians have survived in small groups that operate within well-defined boundaries and in relative isolation. The biological isolation has been maintained by the formation of a unit based on marriages, and this formation has been maintained by the cultural identity associated with the language spoken. Melanesians, especially speakers of Non-Austronesian (Papuan) languages, are characterized by linguistic diversity; Wurm (1982) estimated the total number of Non-Austronesian languages at about 750. Owing to the fact that there are approximately three million speakers in total, the number of speakers per language is only 4000. The Gidra are one such group since they consist of 1850 people (in 1980) in 13 villages and occupy 4000 km² in lowland Papua. Furthermore, the environment diversity among the village (sub-population) groups attracts our concern in terms of population ecology.

THE GIDRA AND THEIR HABITAT

The Gidra-speaking people inhabit the eastern part of the deltaic lowland, called Oriomo Plateau, between the Fly River in the north and the Torres Strait in the south (see frontispiece). Geomorphologically, the plateau is characterized by a swampy plain, with a low ridge, less than 60 m above sea level in the highest part, extending west to east in the central zone. The meandering rivers run south to the Torres Strait or north to the Fly River, and possess a number of tributaries which form a complicated network of creeks in the inland area. According to meteorological records in Daru, where the nearest observatory is located, average monthly temperature fluctuates only from 25°C to 28°C (McAlpine, 1971; McAlpine et al., 1983). With great year-to-year variation, annual rainfall averages about 2000 mm, 80% of it falling in the wet season from December to May. In the wet season the rivers and creeks frequently overflow, while in the dry season the creeks form segregated waterpools and occasionally dried depressions.

The dominant vegetation type in the Gidraland is monsoon forest, while man-induced *Melaleuca* savanna develops mainly along the central ridge. The area is rich in fauna, compared to other parts of the island of New Guinea (Paijmans et al., 1971; Dwyer, 1983). Major game animals are the agile wallaby (*Walabia agilis*) in the savanna, and pigs (*Sus scrofa*), cassowaries (*Casuarius casuarius*), and small-sized wallabies (*Dorcopsis veterum* and *Tylogale* sp.) in the woodland; deer (*Cervus timorensis*) were newly introduced to the area in the 1960s, and their numbers have increased since the 1970s (Downs, 1972). A variety of fishes, shellfishes, and crustaceans are also harvested from the rivers and the sea.

Nothing is known about the situation in the Gidraland several centuries ago, but prehistorians suppose that the cultural identities of the human groups on the southern lowland of New Guinea, in general, might have emerged only in the last 300 years (Swadling, 1983; Hope *et al.*, 1983). Early documents of the Gidra were written, in sketchy fashion, by government patrol officers just after the turn of the present century. From these patrol reports and some ethnographic work among the groups surrounding the Gidra (Beaver, 1920; Landtman, 1927; Williams, 1936), we have learned that the basic structure of their present subsistence pattern seems to have been established a long time ago, although the villages might have been smaller in size and less permanent than at present.

The Gidra language was identified as one of the four languages of the Eastern Trans-Fly Family, belonging to the Trans-Fly Stock (Wurm, 1971, 1982). According to the lexicostatistical test using a 200-item vocabulary list, 40% or fewer vocabulary items are common to each pair of these four Eastern Trans-Fly languages (Wurm, 1971; Fleischmann and Turpeinen, 1976); this implies the difficulty of verbal communication even between people of the same linguistic family.

The Gidra group numbered 1850 in 13 villages in 1980, when we conducted a detailed investigation to make a list of the inhabitants. Their territory was estimated at about 4000 km², and thus their population density was less than 0.5 per km². This low population density, a figure which is more common among hunting-gathering groups than among horticultural groups in the tropics, is one of the important aspects of the Gidra's ecological conditions.

The Gidra group does not have a traditional chief or any political organization governing all of the members. However, they have developed an identity among themselves, which is associated with not only linguistic unity but also sociocultural unity in a broader sense. In particular, clan organization is essential for the Gidra people. There are about 40 totemic clans, and they are classified into two groups, or moieties. Under the patrilineal system, infants of either sex belong to their father's clan and are never transferred even if adopted by other clan members. Sago and coconut stands, the most important properties of the Gidra, are owned by married males and are passed on from a father to his sons at the time of their marriage.

Marriages, traditionally based on the sister exchange rule, take place between men and women who belong to clans of different moieties. In recent years, however, some marriages are organized by payment of a bride price, which may or may not follow the exogamous moiety system. More than half of marriages take place between men and women from the same village. In cases of spouses from different villages, virilocal residence is predominant, although uxorilocal residence also occurs. Each clan possesses a portion of the Gidraland, and boundaries between clan-lands are established throughout the territory. However, each village is inhabited by people—both males and females—belonging to different clans; any villager can exploit the land around the village to obtain natural resources and make gardens, and can even plant sago and coconut palms with the permission of the landowning clan members.

The traditional age-grade system (see frontispiece) is still maintained. A baby is named several months after birth, and he/she is then recognized as a formal member of the society, entering the first age-grade, called *sobijogbuga* for a boy and *sobijogngamugai* for a girl. The entrance to the second age-grade, *yambuga* for boys or *ngamugaibuga* for girls, is based on the acquisition of verbal communication skills. The age at changing of grades was estimated at about eight (seven to nine) years for both sexes; estimation of their age was based on various factors such as birth order of the villagers and our records of birth dates of several villagers, and their age-grades in the follow-up survey period. The girls normally marry at the age of 17–19 years, when they enter the *kongajog* age-grade. When a *yambuga* boy becomes 16–17 years old and his physical maturity is recognized by the married males, he enters the *kewalbuga* age-grade. In their late twenties *kewalbuga* males usually marry and enter the *rugajog* grade. There are two more age-grades, *nanyuruga* and *miid*, for males, and one, *nanyukonga*, for females. Among the age-grades of the married, the boundaries are not clearly explained by the villagers, and their social functions do not markedly differ, although the age at which women become *nanyukonga* roughly corresponds to that of menopause. The age-grade system is used in the bulk of our analyses to distinguish the people by age, since it is still difficult to know their exact ages.

Monogamous families and polygynous families coexist in the Gidra society, the former being predominant. The first marriage of a male leads to the establishment of a new household. The basic unit of production and consumption of the Gidra group is the family or household. The Gidra people's food-getting activities, except for group hunting, can theoretically be completed by the labor force within a single household—a husband and his wife or wives—based on sexual division of labor. However, two conditions are usually not met for this labor system (Morren, 1974). First, aged men tend to retire from such activities as hunting, cutting sago palms, and climbing coconut palms, and therefore cannot perform all the tasks of the male labors. The second condition is the lack of possession of a sufficient number of sago and coconut palms, both of which reach mature size after about ten years. These palms are inherited through the male line, but younger couples usually own a small number of these stands. Inter-household organization of sago-making parties or coconut-fetching parties and a distribution system for game animals contribute to subsistence maintenance in all households of the village. Nonetheless, such cooperation and mutual aid between households do not function beyond a single village. Thus, the village is recognized as an independent unit for subsistence (production and consumption); in terms of long-term population reproduction, however, a single village group does not function as a unit of adaptation since marriages cannot be completed within it.

During our survey period, the people inhabited 13 villages, the adjacent ones 5–20 km apart (see frontispiece). The village sites were changed about every ten years due to the deterioration of the dwellings. When unexpected accidents such as the deaths of several people occurred, the village would also be moved promptly. In recent years, however, political and/or economic factors played significant roles in some of the village transfers. For instance, Ume villagers changed their village site in 1985 because of a conflict over landholding with an adjacent linguistic group. The distance between the old and new villages varied from less than 1 km to about 5 km.

The 13 Gidra villages differ from one locality to another (see frontispiece). Only Dorogori is located along the coast; this village was previously located several kilometers south of the present village site of Abam, until about half a century ago. Abam, Woigi (formerly called Peawa), and Wuroi (Zim) are located along the Oriomo River, and Ume along the Binaturi River. Six villages—Kuru, Wonie, Gamaeve, Podare, Wipim, and Iamega—are inland, at a distance from the permanently running river. Kapal and Rual villages in the northernmost part are located close to the Bituri River. Thus, the 13 villages can be categorized into coastal, riverine, inland, and northern groups. The village locality largely determines the subsistence pattern on the one hand and reflects the degree of modernization on the other. This locality-based categorization of the villages was applied to our research strategies.

The food-getting activities of the Gidra people involve collecting of wild resources, exploitation of *Metroxylon* sago, exploitation of coconuts, slash-and-burn horticulture for plant foods, and hunting and fishing for animal foods. One striking characteristic is the lack of animal husbandry, despite the fact that pig raising is very popular throughout Papua New Guinea. As is discussed in Chapters 1 and 9 in detail, sago production is more important in the inland and northern villages than in the riverine and particularly the coastal villages, because sago palms seem to be unable to tolerate high water salinity in the environment. The riverine and coastal villagers depend on horticulture more than do the inland and northern villagers. For animal food procurement, hunting plays an almost exclusive role in the inland villages, while fishing tends to take the place of hunting in the other village groups.

The center of modernizing influence on the Gidra people has been Daru, the capital of Western Province, with a population of about 7000 in 1980. In order of their distance from or the convenience of transportation to Daru, the coastal village (Dorogori) has been the most modernized, followed by the riverine villages and then the inland and northern villages. Dorogori villagers frequently visit Daru, traveling one or two hours in sailing canoes. The riverine villagers have possessed at least one engine-powered canoe for the trip to Daru since the 1960s. In contrast, the people of the inland and northern villages have to walk long distances to riverine villages to take passage in canoes. The major purposes of visits to Daru are to sell local products (garden crops, coconut, sago flour, game meat, etc.) at the local market and to purchase imported foods (rice, wheat flour, tinned fish, etc.) in the supermarkets.

Though the Gidra people are still following a subsistence regimen, various aspects of their way of living have been influenced by modernization. Although, as mentioned above, inter-village differences in the speed of modernization are essential for the Gidra people's adaptive mechanisms, several significant mechanisms will be delineated here for the Gidra group as a whole. Iron axes and bush-knives were introduced long ago; the aged villagers say that they used those implements during their adolescence. Consequently, the people do not use stone implements for cutting trees in new garden sites or for felling sago trees. Other apparatuses which brought significant technological changes in their food-getting activities were shotguns and fishing nets, although these were introduced in the 1960s and were still owned by a few villagers during our study periods. The only cash crops of the Gidra people are rubber and chilis, both of which were introduced in the 1970s by government agricultural officers; the production of these crops was still small, in terms of cash earnings, in the early 1980s. The agricultural officers also introduced some kinds of garden crops, such as new varieties of sweet potato, although such introduced strains may have been highly productive only for the first one or two years.

The introduction of primary school education and of Christianity influenced the people's behavioral pattern and way of thinking. Both the primary school and the church were first established in the 1960s in the area. At present, most boys and girls can attend a school in their own or a nearby village, and each village has a church, with a missionary or a pastor who was born in the same village or one nearby. One of the prominent changes is seen in the weekly cycle of their behavior. The people, as a rule, do not work on Sundays, and most of the boys and girls who attend school stay in the dormitories on weekdays and come back to the village to stay with their parents from Friday afternoon to Sunday midday. Thanks to school education, the people, especially the younger ones, have come to understand English, and information can reach them through broadcasts in English from Daru and Port Moresby, the national capital. In addition, increased contact with government officers and other people either in or out of the Gidraland has broadened the people's knowledge. These factors make it easy for the villagers to migrate to urban areas. Medical services were available only in Daru before the 1960s, but since then several aid posts have been built in the Gidraland, and a health center was opened in Wipim village in the 1970s. A supply of anti-malarial tablets and immunization, in particular, have contributed to betterment of their health and longevity.

THE RESEARCH PROCEDURE

Following a short-term reconnaissance survey in the eastern part of the Oriomo Plateau in 1967, one of us (R.O.) conducted an eight-month survey in a single Gidra-speaking village, Wonie, from June 1971 to March 1972, focusing upon the adaptative mechanisms of *Metroxylon* sago starch-dependence (Ohtsuka, 1983). In the first several weeks of this period, Dr. Hitoshi Watanabe carried out a functional-ecological survey of the people's hunting weapons in the same village (Watanabe, 1975).

Nine years later, in 1980, two of us (R.O. and T.K.) visited the Gidraland for three and a half months, with three major purposes. First, we visited all the Gidra-speaking villages to explain the purpose and contents of our planned survey in 1981–82 to the people. Second, we selected four villages to be intensively investigated in 1981–82. Our third purpose was to obtain some basic data on the genealogies and the stature and body weight measurements of all inhabitants, which greatly helped us to prepare the research strategy for the main survey.

Five of us (T.S., T.A., and T.I. in addition to the above two) cooperated in the field for six months from July 1981 to January 1982. The investigations in this period were divided into two major series. First, four of us (R.O., T.A., T.K., and T.I.) stayed in the four selected villages: Rual in the north, Wonie inland, Ume in the riverine area, and Dorogori in the coastal region. One of us worked in each village, engaged in joint time allocation and food consumption surveys and in other independent surveys according to the interest of each of us. Second, several short-term patrol surveys by two to four members of the team were conducted to obtain anthropometric, physiological, and demographic data from all of the 13 villages, and intensively from the four villages. In addition, samples of scalp hair, urine, food, and drinking water were collected in various locations throughout the research period and later analyzed in laboratories in Japan, in cooperation with several scholars, particularly T.H., one of the contributors to this book.

Since 1985, follow-up studies among the Gidra have been done three times, and some of the results obtained from the four intensively studied villages in 1986 are presented in this volume.

(*Ryutaro Ohtsuka and Tsuguyoshi Suzuki*)

I. THE ECOLOGY OF FOOD PRODUCTION

Overview

In human ecology studies, the production and consumption of food has been viewed as a key component of adaptive mechanisms; quantified data on food production and consumption have provided insights on how people adapt to their environment. From the standpoint of human population ecology, special attention is paid to the differences in food-centered activities among village groups (subpopulations). Under the heading "The Ecology of Food Production," we assemble six papers which look at the quantitative aspects of the Gidra food production in relation to biological and behavioral features, largely at the expense of the qualitative aspects such as the cognitive, social, and technological; the latter aspects are mentioned in R.O.'s book, *Oriomo Papuans: Ecology of Sago-Eaters in Lowland Papua* (Ohtsuka, 1983).

Time allocation data (Chapter 1) show the kinds of food procurement activities undertaken by the Gidra people, and to what degrees. Biobehaviorally, individual-based work efficiency is of great concern, and in fact age-dependent and independent individual variations play crucial roles in hunting productivity (Chapter 4). Individual sensorimotor functions, e.g. visual acuity (Chapter 2) and grip strength (Chapter 3), are studied as related physiological variables, while, at the same time, these functions for individuals of both sexes and various ages identify and characterize them in the wider biological as well as the social organizational sense. For plant-food-getting activities, work efficiency differs to a lesser degree among individual workers, so their work efficiencies are judged in relation to the botanical natures of the target plants and to the work procedures (Chapter 5). Measurement of energy expenditure makes it possible to evaluate work loads of various food-getting activities and contributes to the overall energy input-output analysis (Chapter 6).

Chapter 1

Time Allocation for Food Procurement

As a basic tool for elucidating the structural aspect of human behavioral adaptation, time allocation study has been developed and widespread in anthropology, human ecology, and their related fields (e.g. Watanabe, 1977; Gross, 1984). Following the pioneer works on human labor time (e.g. Richards, 1939; Erasmus, 1955; Conklin, 1957), the random spot-check method used by Johnson (1975) among the Machiguenga horticultural group has improved the precision of data collection, and has had major influence on the researchers in these fields (e.g. Munroe *et al.*, 1983). For instance, the use of binary code, based on physical description and description of consequence for the subject's behaviors on randomly selected occasions, which was proposed by Mulder and Caro (1985), is suggestive of the validity of time allocation data for cross-cultural comparison. As mentioned in my comments on their article (Ohtsuka, 1985), however, the applicability of any time-allocation method differs according to the local environmental conditions of the target group.

Among the Gidra people, who inhabit a densely wooded environment and whose work places are scattered over a wide area, the random spot-check method is not appropriate, simply because, on the one hand, it is too difficult for the researcher to move around the subjects' working spots, and, on the other hand, the subjects are almost always able to perceive the presence of the researcher before he comes within sight in their working places and thus tend to change their behavior before the time of encounter. In our study, each villager's time spent on subsistence activities was estimated by recording his/her time of departure from and return to the village; we also did activity-tracing surveys on many separate occasions to understand in detail what the people did and how in various working places (Ohtsuka, 1977b, 1983). The time-recording survey was conducted in four villages by one of our research team members for successive 14 days in 1981: in Rual by T.K., in Wonie by myself, in Ume by T.A., and in Dorogori by T.I. The same survey had been done for 13 successive days in 1971–72 in Wonie by me in both the dry and wet seasons.

The present chapter assembles the data from the time-recording survey in the four villages in 1981 and in Wonie during the dry season of the 1971–72 survey period, since our 1981 data were collected only in the dry season. The basic purpose of this chapter is to clarify what subsistence activities the Gidra people's time was spent in and for how long, and, in particular, to make an inter-village comparison.

DATA COLLECTION AND ANALYSIS

Throughout the period of the time-recording survey, a researcher checked the time of departure from and return to the village from morning till evening for all adult villagers (including *kewalbuga* for males and elder *ngamugaibuga*, or over approximately 15 years old, for females), and asked them upon their return where they had been and what they had done, with help of one or two assistants from the same village, who watched the people's movements in different spots within the village settlement. To do this study, it was necessary for each researcher to become fully acquainted with the people and the place names, about 200 in number, within each village-land of 50–100 km².

The observed time duration in the day was 14 hr, from 6:00 to 20:00. Time calculated as spent in the activities included that spent walking back and forth between the village and the working site. In cases which involved overnight visits to other villages or to Daru town, the hours spent outside the village were omitted from the analysis. In other cases, including overnight activities within the village-land, generally during sago-making trips, an estimation was made to calculate the time. According to our observations in activity-tracing surveys of such cases, the average proportion of time spent at the working site to the total time between 6:00 and 20:00 was 60%, so this proportion was applied to the estimation of actual labor time during out-of-village hours.

Our selection of the subject villagers resulted in the omission of children's activities. This came primarily from the difficulty of pursuing the children's frequently changing activities, although whenever they followed adult villagers, their time was also checked. However, our activity-tracing survey disclosed no significant contribution to subsistence production by the children, if such non-food-producing activities as caring for younger siblings were excluded. The second issue in our methodology was related to the duration of the single session of time-recording, that is, 14 or 13 days. This investigation period was determined by our preparatory observations, which demonstrated that animal-getting activities like hunting and fishing were practiced without any manifest cycles but usually at least once within a week or so by the individuals who were actively engaged in such activities, and that sago-making tended to be practiced every ten days or two weeks by each household member. Judging from the results, the setting of this investigation period met with success, except in a few cases, as will be mentioned later.

The final problem was how to categorize various subsistence activities for inter-village comparison. In this analysis, they were classified into six major types, i.e., hunting, fishing, collecting, sago exploitation, coconut exploitation, and horticulture (including time spent in fetching firewood, which was normally indistinguishable from horticultural activity), despite the fact that not only the repertoire of activities but also the content of activities belonging to the same category (e.g. rod-line fishing in the creeks, spearing in the rivers, and netting in the sea within "fishing" category) differed from village to village. It is also noticed here that the present analysis neglected some activities which did not directly procure foods but were related to their subsistence or daily living: for instance, building or repairing houses, making canoes and hunting weapons, and fetching drinking water from nearby waterplaces. However, traveling to Daru to sell local products and participating in wage labor were recognized to contribute to the people's sustenance to the similar extent as the food-getting activities *per se*. In fact, during the survey period these activities were observed only in Dorogori.

INTER-VILLAGE COMPARISON OF LABOR TIME

Based on the total time spent on each activity (type) by each individual for the whole observation duration, the per-day per-person time broken down by sex and by village is shown in Table 1. Of these data sets, that for Rual villagers seemed not to represent their activity patterns for long-term (e.g. one month) periods. The most critical difference was that a large proportion of this village people's time was usually spent in sago exploitation according to our (particularly T.K.'s) observation, but in the survey period they scarcely went to sago groves since their sago-making was accidentally concentrated just before and/or just after this period; instead, their time might be used for hunting and fishing a little longer than usual. In another example, as mentioned above, Dorogori villagers were engaged in visits to Daru to sell the local products and in wage labor for road construction. The former was regularly practiced, while the latter coincidentally took place in the survey period. In fact, the average per-day time spent in travel to Daru was 46 min for males and 41 min for females, and that in wage labor was 45 and 29 min, respectively; the total work time spent in both these activities and subsistence activities is also mentioned in Table 1.

Table 1. Per-Day Time (in Minutes) Spent in Food-Getting Activities, Broken Down by Village and by Sex

Group (N)		Hunt-ing	Fish-ing	Collect-ing	Sago	Horti-culture	Coconut	Total[a]
Rual: 1981								
male	(22)	138	11	16	13	45	0	223
female	(36)	0	43	17	19	87	0	166
both	(58)	52	31	17	17	71	0	188
Wonie: 1971								
male	(23)	76	6	4	105	88	10	289
female	(18)	0	0	1	171	120	21	313
both	(41)	43	3	3	134	102	15	298
Wonie: 1981								
male	(28)	46	1	7	101	123	0	278
female	(32)	0	6	9	127	138	0	280
both	(60)	22	3	8	115	131	0	279
Ume: 1981								
male	(45)	24	27	7	49	69	8	184
female	(43)	0	7	1	59	220	1	288
both	(88)	12	17	4	54	140	5	232
Dorogori: 1981								
male	(29)	15	6	1	0	76	6	104 (195)
female	(49)	0	23	0	0	88	4	116 (186)
both	(78)	6	17	0	0	84	5	112 (189)

[a] Figures in parentheses for Dorogori villagers' total time involve time spent in wage labor and travel to Daru.

With these reservations, the results show that time spent in subsistence activities markedly differed among the village groups in quality and in quantity. The total per-day work time for all subjects (of both sexes) was longest among Wonie villagers (nearly 5 hr in both survey periods), followed by Ume villagers (about 4 hr), and then the remaining two villagers (about 3 hr). For the difference between Wonie and Ume villagers, a most manifest reason came from the labor system for horticulture. In Wonie, where gardens were usually made by each household, a husband and his wife (or wives) formed a basic working unit, whereas in Ume, where the gardens were mostly made by cooperative efforts of several households, women tended to go to the gardens by themselves: this tendency was reflected in shorter horticultural time of males in Ume. When compared to other societies, the Gidra's labor time seems not to be longer: using "outside the house" activities (directly pertaining to food production) in Minge-Klevana's (1980) comparative analysis, even the longest time of Wonie villagers, i.e. 5 hr, is placed in the "shorter-working" group; daily work time of 3–5 hr of the Gidra is rather comparable with that of hunting-gathering groups, e.g. the Kalahari San (Tanaka, 1980) and Arnhem Land Australians (McCarthy and McArthur, 1960).

There are three more observations in the figures in Table 1. First, the ratio of the time spent in sago exploitation to that in horticulture in each village group, except Rual, was fairly identical with the ratio of the consumed amount (in weight or in food energy) of sago starch to that of garden crops, which were investigated for 12 or 14 successive days on different occasions in the dry season of the same year (the data are shown in Chapter 9 in this volume); sago starch consumed by Dorogori villagers was, however, not provided by their labor but mostly purchased from the local market in Daru with the earnings from selling such local products as coconuts and garden crops. Second, the ratio of time spent for hunting to that for fishing in each village (including Rual) was also similar to the ratio of the consumed amount of game animals to that of aquatic animals. Finally, the comparison between the data from Wonie in two periods over a ten-year interval demonstrates that the subsistence pattern has basically been identical but has changed, for instance, in the disappearance of coconut exploitation in 1981, and the alternation in the order of labor time length between sago exploitation and horticulture. The former reflected the increased predation of coconut fruits by giant rats (*Rattus rattus*), as a consequence of which there was no non-damaged mature fruit in 1981. The latter was perhaps derived from the introduction of high-yielding varieties of sweet potato in that duration on the one hand, and, on the other, from the people's dislike of painstaking sago-making labor in a swampy environment.

SOME FURTHER ANALYSES

In our understanding, the time allocation data can be well evaluated when related to other aspects such as use of space, activity rhythms, labor efficiency, energy input-output ratio, and organization of working groups (Ohtsuka, 1977b, 1983); some aspects are discussed in other chapters of this volume. In this chapter, labor time by sex/age groups and labor efficiency by plant-food-getting activities will be compared.

Work time varies in quality and in quantity between sex and through aging. Here, the subjects were classified into six groups based on the Gidra age-grade system: *nanyuruga* (including *miid*), *rugajog*, and *kewalbuga* for males, and *nanyukonga, kongajog,* and *ngamugaibuga* (the elders only) for females. This analysis treated three major activity categories: animal-food-getting activities (i.e. hunting and fishing combined), sago-making, and horticulture, the latter two of which were major plant-food-getting activities. Figure 1 shows per-day per-person time spent in these three

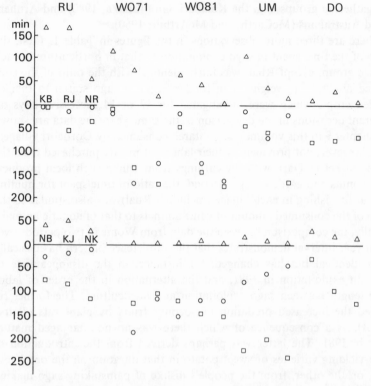

Fig. 1. Per-day per-person time spent in three activity categories by sex/age group and by village.
RU: Rual in 1981, WO71: Wonie in 1971, WO81: Wonie in 1981, UM: Ume in 1981, DO: Dorogori in 1981. △ Hunting/Fishing, ○ Sago exploitation, □ Horticulture.
KB: *Kewalbuga,* RJ: *Rugajog,* NR: *Nanyuruga,* NB: *Ngamugaibuga,* KJ: *Kongjog,* NK: *Nanyukonga.*

activity categories by sex/age group and by village; however, there were no elder *ngamugaibuga* girls in Wonie in 1971, and Dorogori villagers did not practice sago-making.

The sex and age differences in labor time can be summarized into three aspects. First, animal-food-getting activities were mainly done by males (see also Table 1); Dorogori villagers did not show this tendency largely because the labor for gill netting, which was the most popular fishing method in this village, was practiced by women more than men. Second, sago-making and horticulture were practiced by women more than by men, although the sexual difference was small compared to animal-getting activities. Third, the changing time use pattern for subsistence

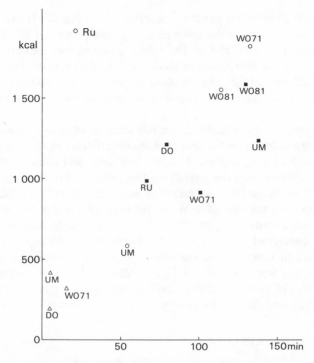

Fig. 2. The relationship between per-day per-person time and per-day per-adult male energy intake for major plant-food-getting activities.
The indications attached to the plots are same as in Fig. 1. ○ Sago, △ Coconut, ■ Horticulture.

efforts by aging differed between the sexes. Except for Dorogori villagers, men's time was characterized by decrease in animal-getting activities with aging; plant-getting time of unmarried *kewalbuga* males was particularly short. To contrast, women's time for any activity was relatively identical throughout their lives. These characteristics are important ecological conditions on labor organization of the Gidra people, and perhaps applicable to many other societies.

Labor efficiency, in terms of food energy produced per unit of labor time, can reasonably be evaluated for plant-food-getting activities, since plant foods act mainly as energy sources. Based on our above-mentioned food consumption records, per-day energy intake adjusted to per-adult male value was broken down according to the source food groups, each of which corresponded to subsistence activity type or purchasing activity (Chapter 9 in this volume). Here, per-day per-person time spent in each of the three activities and per-day per-adult male energy intake from foods procured by it are plotted in a scattergram (Fig. 2). It can be said that the results show similar labor efficiency, irrespective of the kinds of food-getting strategies and of the villages, except for sago-making by Rual villagers; the correlation analysis for eight data sets of sago-making and horticulture (except sago-making in Rual) gave $r = 0.805$ (p < 0.05) and that for 11 data sets involving coconut-fetching, $r = 0.926$ (p < 0.01).

In this connection, a higher labor efficiency in sago exploitation than in horticulture, based on the results of Wonie villagers in 1971, is stressed in Chapter 5 of this volume. This was the case, and such condition is not contradictory with the overall tendency observed in the four villages in 1981. Nonetheless, it is necessary to assess the diversity of labor efficiency, including the change in Wonie between 1971 and 1981, taking the local conditions into account. The most distinguishable features for the variation concerned derive from lower horticultural efficiency in Wonie in 1971 and in Ume than in the others. In my judgement, the possibly related factors were the lack of high-yielding varieties of garden crops (particularly, of sweet potato) in 1971 for the former (see Chapter 9 in this volume), and the low proportion of men's labor input for the latter.

CONCLUDING REMARKS

Methodologically speaking, our data based on the time-recording method may be regarded as crude indicators for the people's time allocation. However, it can be judged that the results represent the overall differences in activity pattern among the village groups and between two different time periods in the same village, Wonie. Moreover, the results seem meaningful in evaluating the infrastructure of the local adaptation. In general terms, time allocation data are capable of providing one of the basic sets of information on the people's adaptive mechanisms. In other words, such data are indispensable, but can fully illuminate the adaptive mechanisms when related to other aspects, taking the minute local conditions into account. This aspect is particularly important among the Gidra group, whose adaptive system differs from village to village and has been rapidly changing.

(*Ryutaro Ohtsuka*)

Chapter 2

Visual Acuity as a Sensory Function

Keen vision, in the sense of true visual acuity as well as learned ability to recognize visual clues, is indispensable to people dependent on hunting activity, and the reported keen vision of hunters and gatherers seems to have a firm physiological basis (Mann, 1957; Neel *et al.*, 1964). Several investigations of refractive aberration show that nonliterate peoples are comparatively free from astigmatism and myopia, while civilized peoples are frequently characterized by high percentages of the shortsighted and astigmatic (Kalmus, 1969). Indeed, from an evolutionary perspective, Post (1962) suggested that the inter-group differences of vision could be attributed to "a positive selection favouring low grades of myopia, and a relaxation of selection for acuity which would permit the accumulation in the population of a wide variety of hereditary deficiencies."

In our 1981–82 survey period, visual acuity was tested and the anterior portion of the eye was inspected in four Gidra villages. This chapter reports on the Gidra people's visual acuity and examines the relationship between visual impairment and the degree of cataract (or corneal opacity), and documents the diminution in acuity with age in relation to hunting activities and to the Gidra traditional age-grade system.

SUBJECTS AND METHODS

Tests of visual acuity and inspection of the anterior portion of the eye were carried out in July and August 1981 in four selected villages: Rual, Wonie, Ume, and Dorogori. Three hundred fifty-six persons (177 males and 179 females) were examined, including two persons blind in one eye: a boy who had been blind in his left eye from birth and a man who was blinded in his right eye in childhood. Children under approximately ten years of age were excluded from the examination on account of their inability to participate.

Subjects were categorized into three age groups according to the age-grade system: (1) youth, i.e. *kewalbuga* for male and *ngamugaibuga* for female, (2) adult, i.e. *rugajog* for male and *kongajog* for female, and (3) elder, i.e. *nanyuruga* and *miid* for male and *nanyukonga* for female. Estimated age ranges of the three groups are: youth, 10–20 years old; adult, 20–50 years; elder, over 50 years.

Visual acuity varies with external conditions like illumination and test objects as well as internal conditions like motivation, attention, and intelligence. The method of examination was as standardized and objective as possible under field conditions. Using the international Landolt rings test chart in daylight in the open air, each eye was tested separately. The chart was hung on a bright wall avoiding the direct rays of the sun and without any electrical illumination, since no Gidra village was supplied with electricity. In order to avoid errors caused by communication difficulties, the procedure of examination was explained to each subject before the test by a local assistant. A subject standing at a distance of 5 m from the test chart was asked to point to a break in a Landolt ring. When he/she failed to recognize the largest symbol (acuity of 0.1, i.e. 20/200), he/she was brought closer to it, and the distance at which he/she recognized it was recorded. All the subjects were intelligent enough to grasp the instructions, responding quickly and indicating the break in the ring with a firm gesture.

Visual acuity is designated by a decimal notation which represents the reciprocal of the visual angle in minutes, while it is usually expressed in terms of Snellen's fraction given in feet or meters; for example, 20/20 in feet (or 6/6 in metres) = 1.0; and 20/200 (6/60) = 0.1 (Mittelman, 1980; Newell, 1982).

The anterior portion of the eye of adults and elders was inspected without any mydriatic treatment. The degree of opacity present was classified into three categories: negative (no opacity in lens or cornea), slight (opacity only peripherally), and dominant (opacity in most of the lens and cornea).

RESULTS

The results of the visual acuity test for the better eye are shown in Figs. 1 and 2, for each sex. In the majority the acuity did not differ between right and left eyes (Table 1), although in two cases one eye had been damaged. The frequency distribution of the acuity is obviously asymmetrical in both sexes, with much the highest frequency in the largest class (2.0, the maximal value of the test) and a long thin tail to minimal acuity. Medians of males and females in the better eye were respectively 2.0 and 1.5, and the corresponding means (SDs) were 1.55 (0.64) and 1.44 (0.64). The proportion of normal vision (acuity of 1.0 or better) to the total

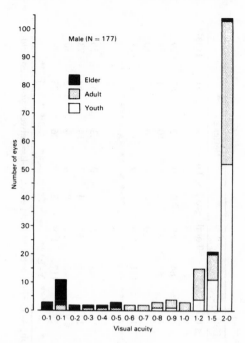

Fig. 1. Visual acuity of the better eye of males.

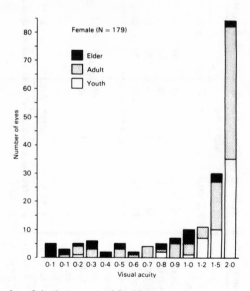

Fig. 2. Visual acuity of the better eye of females.

Table 1. Visual Acuity by the Three Age Groups with Results of χ^2-Test

	Right eye					Left eye					Better eye				
	≤0.2	≤0.6	≤1.0	≤2.0	Total	≤0.2	≤0.6	≤1.0	≤2.0	Total	≤0.2	≤0.6	≤1.0	≤2.0	Total
Male															
Youth	0	1	3	65	69	0	0	4	64	68	0	0	2	67	69
Adult	3	11	15	58	87	5	4	14	65	88	2	5	10	71	88
Elder	14	4	1	1	20	15	3	0	2	20	14	4	0	2	10
Total	17	16	19	124	176	20	7	18	131	176	16	9	12	140	177
χ^2-test[a]	T*** Y-A*** A-E*** Y-E***					T*** Y-A** A-E*** Y-E***					T*** Y-A* A-E*** Y-E***				
Female															
Youth	1	1	8	46	56	1	0	6	49	56	1	0	3	52	56
Adult	5	9	13	66	93	7	7	14	65	93	4	7	14	68	93
Elder	10	8	7	5	30	9	9	9	3	30	8	8	9	5	30
Total	16	18	28	117	179	17	16	29	117	179	13	15	26	125	179
χ^2-test[a]	T*** Y-A A-E*** Y-E***					T*** Y-A* A-E*** Y-E***					T*** Y-A* A-E*** Y-E***				

[a] T = between the three age groups, Y-A = between youth and adult, A-E = between adult and elder, Y-E = between youth and elder.
* $p < 0.05$, ** $p < 0.01$, *** $p < 0.001$.

number of eyes for all subjects was 74% (528 of 710 eyes); taking only the better eye into account, the percentage rises to 78% (81% for males and 75% for females).

Figures 1 and 2 show that the youths have the best visual acuity and the elders the poorest. In Table 1 the values of visual acuity are classified into four grades. Ninety-seven percent of boys (67 of 69) and 93% of girls (52 of 56) exceeded 1.2 vision in the better eye. Adults largely retained this high visual acuity, for 81% of males and 73% of females had an acuity of 1.2 or better; however, the difference in visual acuity between youths and adults is statistically significant. For youths and adults combined, 88% of males and 81% of females exceeded a value of 1.2. By contrast, visual acuity of elders was considerably lower than that of adults. Visual reduction in females is less than in males; eighteen of the 20 male elders had an acuity lower than 0.6, while about half of female elders (14 of 30) maintained vision of 0.7 or better. Chi-squared tests between the three age groups demonstrate a significant decrease in visual acuity with age in both sexes (Table 1).

Figure 3 shows the number and the proportion of cases of cataract (or corneal opacity) in adults and elders, judged by inspection. Opacity (slight or dominant) was found in 28% (30 of 108) of men and 21% (26 of 123) of women. Cataract (or corneal opacity) was significantly more prevalent in elders than in adults of both sexes, and rather more so in males: 75% of male elders and 70% of female elders; 17% of male adults and 5% of female adults (Fig. 3).

The existence of cataract (or corneal opacity), even when slight, impairs the visual acuity in both men and women (Table 2).

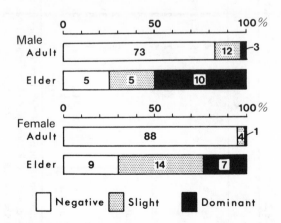

Fig. 3. Number and proportion of cataract (or corneal opacity) of adults and elders; for classification see text.

Table 2. Visual Acuity with or without Cataract (or Corneal Opacity)

| | Visual acuity | | | | | | | | | | | | | | |
	<0.1	0.1	0.2	0.3	0.4	0.5	0.6	0.7	0.8	0.9	1.0	1.2	1.5	2.0	Total
Male															
Negative		4 (5)	1 (1)		3	1 (1)	4	4 (2)	2 (1)	2 (2)	3 (5)	6 (9)	11 (13)	36 (37)	77 (78)
Slight	2 (1)	1 (2)	1 (2)	(1)	1		1	(1)	1	2	(2)	2 (1)	2 (1)	3 (2)	17 (17)
Dominant	2 (2)	5 (5)	1 (2)	2 (1)	1	(1)	1 (1)	1	(1)	1			(4)	1 (2)	13 (13)
Female															
Negative		2 (1)	2 (3)	3 (5)	(1)	5	2 (3)	1 (2)	3 (2)	3 (6)	7 (7)	9 (10)	25 (22)	35 (35)	97 (97)
Slight	2 (2)	1	3 (5)	1 (3)	1 (1)	2 (1)			1 (3)	2	1 (2)		1 (1)		18 (18)
Dominant	3 (3)	2 (2)		(2)	1	1 (1)	(1)	1							8 (8)

Table shows number of right eyes (left eyes in parentheses).
According to Kendall's rank correlation coefficient: −0.450 (p<0.001) for male right eye; −0.441 (p<0.001) for male left eye; −0.524 (p<0.001) for female right eye; −0.506 (p<0.001) for female left eye.

DISCUSSION

The finding that 88% of the males and 81% of the females exceeded 1.2 vision, when elders were excluded, can be compared with results of other studies in hunters and gatherers, although there are only a few that report quantitative data. Among 77 Cayapo Amerindian males of estimated age greater than nine years 71% had better than 1.3 (20/15) vision (Neel *et al.*, 1964); the Xavante, another Amerindian group, had a still more impressive performance, 12 of 13 males aged 15–30 years showing an acuity of 1.3 (20/15) or 2.0 (20/10) (Neel *et al.*, 1964). For an aboriginal Australian population, Mann (1957) reported that of the 146 individuals of all ages tested, for the sexes combined, 119 (82%) had vision of 1.2 (6/5) or better in the better eye.

For Papua New Guinean populations, there are only two reports on visual acuity, though these in their categorization of normal vision did not distinguish values of 1.0 (6/6) from better ones. Vines (1970) surveyed three regions of Papua New Guinea (highland, mainland, and island), but not the lowland region where the present subjects live. The percentages of subjects with normal vision in the better eye for the sexes combined were 81%, 86%, and 91% respectively among highlanders, mainlanders, and islanders. Sinnett (1975) found that 74% of males and 83% of females of Enga-speaking populations in the highland had normal vision. The percentage of normal vision of the Gidra (78% for all subjects; 81% for males and 75% for females) was slightly lower than those in these reports except for the Enga males.

Cataract of the senile type and corneal opacity, which most likely resulted from trauma or neglected ulceration, were common throughout the highlands, mainland, and islands of Papua New Guinea (Vines, 1970). In the Enga-speaking people with a high prevalence of cataract (18% of males and 10% of females) and corneal opacity (7.7% of males and 2.4% of females), the percentage of normal vision fell progressively with advancing age, and visual impairment was more marked in males than in females (Sinnett, 1975). The present data parallel this result. Cataract or corneal opacity was the most prevalent eye disease among Gidra adults and elders: 28% of males and 21% of females. It is concluded that severity of visual impairment of the elders, especially males, is significantly related to this.

Hunting time and animals killed were surveyed in the Wonie village (Chapter 4 in this volume); according to the records in 1971–72, the hunters of the elder age-grade spent markedly less time on hunting than did the adults and youths, and they actually killed no animals. Adults and youths spent similar time in hunting, yet the labor efficiency of the former (1.06 kg/hr) was more than twice that of the latter. The hunting skill of boys and adolescents who are developing into active hunters has not yet reached the adult's level (Kawabe, 1983). Keen vision is essential for an active hunter, and the adults maintained this although their average acuity was lower than that of the youths. Elder males had far lower acuity, and this visual impairment may partly explain their scanty hunting practice and low productivity.

The change from youth to adult age-grade is socially formalized (i.e. at marriage), related to roles and behavior in Gidra society. In contrast to this, the criteria of change from adults to elders seem ambiguous, the most recognizable indicator as explained by the Gidra people being the proportion of gray hair. However, of the various senescent processes, decrease of sensory performance must also be important because reduction in efficiency of activities, such as hunting, is clear for all to see. It seems, therefore, that deterioration of sensory abilities, as represented by visual acuity, is an important determinant not only of an individual's behavior but also of his/her position in the social system, especially where this is based on age-grades as among the Gidra.

(Toshio Kawabe)

Chapter 3

Grip Strength as a Motor Function

Human activity, nutritional status, physiological ability, and body composition are interrelated. Malina *et al.* (1982) disclosed that grip strength of poorly-nourished Zapotec horticulturalists in rural Mexico was lower than the Western standard, although they also suggested that the strength per unit of body weight did not significantly differ. Grip strength per unit of body weight was then compared among pastoral Turkana in Kenya, Zapotecs, and Caucasoids (Little and Johnson, 1986): the Turkana were linear in physique; their diet was characterized by extremely high protein and low energy intake; and the Turkana women, in contrast to the men, tended to develop arm muscle strength by daily lifting and carrying of infants, water containers, and firewood and milking livestock (Little *et al.*, 1983; Little and Johnson, 1986). As a result, grip strength per unit of body weight was higher in Turkana and Caucasoids than in Zapotecs in females.

This chapter analyzes grip strength and body composition of the Gidra in relation to ecological conditions. Special attention is paid to the difference between two ecologically contrasting villages, one inland and less modernized and the other coastal and more modernized, since this setting makes it possible to assess the influence of ecological factors on biological characteristics.

Both grip strength and body composition were measured in two villages, Wonie inland and Dorogori on the coast, in the period from August to October 1981. Anthropometry was conducted by one of our research members (T.K.) to avoid inter-observer error. The measurements which are analyzed here include stature, body weight, upper arm circumference, and triceps and subscapular skinfold thicknesses. From these measurements, upper arm muscle cross-sectional area and fat-free mass were calculated (see Chapter 11 for details).

For measurements of grip strength, a Smedley dynamometer was used to test the maximum voluntary contraction of the grip flexors. After adjustment of the dynamometer stirrup, each subject in a standing position was instructed to grasp it with a maximum burst of energy. This test

Table 1. Number of Subjects by Sex, Age, and Village

	Age					
	17–19	20–29	30–39	40–49	50+	Total
Wonie						
Male	5	6	9	5	5	30
Female	8	4	7	5	6	30
Dorogori						
Male	4	9	5	9	8	35
Female	6	12	10	4	9	41
Total	23	31	31	23	28	136

was conducted several times for each hand, after a time interval of at least several minutes. In this paper, the preferred hand with the maximum force value was taken as the grip strength.

All villagers in Wonie and Dorogori except children younger than approximately ten years old participated in the anthropometric study. Grip strength was tested in the adolescents and adults, although a few older men and women, presumably more than 65 years old, did not participate because of the difficulty of exerting a full burst of energy. Among the adolescents, this chapter treats only those who were estimated to be 17 years of age or older.

In the Gidra society, it is difficult to know the exact age of individuals, especially those who are older. In this study, however, we gave each subject estimated age in decades, based on birth order among the villagers, which is well recognized, and on fragmentary records of birth years registered at the hospital in Daru or by government officers. I have studied the Wonie villagers since 1971–72 so that the estimation of their ages was relatively easy, and among Dorogori villagers, their relatively modernized situation made it possible. A breakdown of the subjects by sex, age-group, and village is shown in Table 1.

RESULTS

Anthropometric measurements and indices relevant to the present analysis are shown in Table 2, and they demonstrate several interage-group and inter-village differences. First, stature differs by age group. The shorter stature of the younger (17–19 and 20–29) groups, in males in particular and also in Wonie females, seems to be attributable to the fact that the Gidra continue to grow in their early 20s (Chapter 13 in this volume); a similar pattern of prolonged maturation has been reported from other preindustrial populations (e.g. Malcolm, 1970c: Little et al., 1983). Second, males' measurements and indices are characterized by

Table 2. Means (SDs) of Selected Anthropometric Data by Sex, Age, and Village

Group (N)	Stature (cm)	Body weight (kg)	Body mass index (kg/cm² × 10⁴)	Upper arm muscle area (cm²)	Fat-free mass (kg)
Wonie male					
17–19 (5)	161.4 (4.3)	52.8 (5.8)	20.2 (1.3)	45.2 (9.3)	46.0 (4.7)
20–29 (6)	164.9 (4.2)	58.2 (4.3)	21.4 (1.0)	47.9 (9.4)	51.0 (2.4)
30–39 (9)	167.3 (3.5)	58.1 (4.2)	20.7 (1.0)	51.9 (6.5)	51.3 (3.8)
40–49 (5)	166.4 (6.5)	57.2 (5.8)	20.6 (1.1)	51.7 (3.0)	50.6 (5.0)
50+ (5)	162.0 (4.1)	52.7 (4.4)	20.1 (1.1)	45.1 (6.4)	47.4 (2.6)
Total (30)	164.8 (4.8)	56.2 (5.1)	20.6 (1.1)	48.8 (6.6)	49.6 (4.1)
Dorogori male					
17–19 (4)	165.5 (5.9)	57.8 (6.5)	21.1 (1.5)	43.7 (7.7)	49.3 (5.5)
20–29 (9)	167.0 (4.8)	63.5 (6.1)	22.8 (1.8)	52.6 (5.5)	54.2 (4.8)
30–39 (5)	169.4 (3.6)	64.9 (5.0)	22.6 (1.7)	55.1 (5.6)	54.5 (2.9)
40–49 (9)	167.0 (6.2)	64.3 (10.8)	23.0 (2.9)	55.4 (8.1)	54.1 (5.5)
50+ (8)	164.0 (4.8)	54.6 (8.7)	20.2 (2.5)	47.0 (8.5)	47.9 (6.7)
Total (35)	166.5 (5.2)	61.2 (8.7)	22.0 (2.4)	51.4 (8.0)	52.2 (5.8)
Wonie female					
17–19 (8)	152.0 (4.8)	47.3 (5.0)	20.5 (1.9)	32.1 (4.5)	35.5 (2.1)
20–29 (4)	153.5 (6.3)	43.6 (3.0)	18.6 (1.5)	36.0 (3.0)	35.8 (2.1)
30–39 (7)	156.4 (5.4)	47.3 (3.1)	19.4 (2.0)	38.0 (5.7)	37.2 (2.4)
40–49 (5)	154.6 (2.1)	43.2 (3.9)	18.1 (1.6)	36.6 (5.0)	36.7 (1.6)
50+ (6)	155.1 (4.0)	42.6 (3.2)	17.7 (1.2)	36.2 (2.3)	36.7 (6.7)
Total (30)	154.3 (4.7)	45.2 (4.2)	19.0 (1.9)	35.6 (4.7)	36.1 (2.3)
Dorogori female					
17–19 (6)	159.3 (5.3)	58.3 (6.1)	22.9 (1.5)	35.9 (7.0)	41.8 (1.7)
20–29 (12)	159.2 (6.6)	56.9 (6.8)	22.5 (2.8)	39.8 (5.0)	42.0 (3.4)
30–39 (10)	158.8 (4.0)	60.5 (7.5)	24.0 (2.6)	42.3 (6.5)	42.5 (4.0)
40–49 (4)	155.3 (3.6)	55.9 (11.5)	23.3 (5.3)	41.8 (3.0)	40.4 (4.1)
50+ (9)	153.6 (3.6)	52.3 (9.9)	22.2 (3.7)	41.7 (4.7)	38.8 (4.0)
Total (41)	157.5 (5.7)	56.9 (8.2)	22.9 (3.1)	40.4 (5.7)	41.2 (3.7)

higher values in the 20–29, 30–39, and 40–49 age groups than in those aged 17–19 and 50+, while this age-related pattern is less marked in females. Third, all measurements are greater in Dorogori villagers than in Wonie villagers, and this tendency is remarkable in females.

The body mass index (BMI) is a good indicator of body composition as well as obesity (National Institute of Health, 1985). The frequency distribution of BMI (Table 3) reveals that all Wonie villagers have a value of BMI lower than 24, whereas a fairly large number of Dorogori villagers exceed this value. Since the Wonie villagers have preserved the traditional way of living, it is reasonable to judge that the individuals with BMI greater than 24 are more strongly influenced by modernization, and they are classified as a high-BMI group in this analysis. However, this group

Table 3. Frequency Distribution of Each Sex/Village Group by Body Mass Index

	N	<18	18–20	20–22	22–24	24–26	26+
Wonie							
Male	30	0	10	17	3	0	0
			(33.3)	(56.7)	(10.0)		
Female		9	16	2	3	0	0
	30	(30.0)	(53.3)	(6.7)	(10.0)		
Dorogori							
Male	35	2	2	16	9	5	1
		(5.7)	(5.7)	(45.7)	(25.7)	(14.3)	(2.9)
Female	41	1	6	9	10	8	7
		(2.4)	(14.6)	(22.0)	(24.4)	(19.5)	(17.1)

does not correspond to the "obese" group, which is usually defined by a greater BMI value, for example, 27 for males and 25 for females in the United States (National Diabetes Data Group, 1979).

Maximum grip strength is illustrated in Fig. 1, in which the mean value for each age group is plotted at the middle of the estimated age range. Comparable data for the Turkana (Little and Johnson, 1986) and the Japanese (Japanese Ministry of Education, Science and Culture, 1985) are also shown in Fig. 1, although inter-population comparison is done with difficulty due to differences in dynamometer design (Shephard, 1985).

Inter- and intra-population comparison reveals several characteristics. First, in all age groups Dorogori males and females, respectively, have

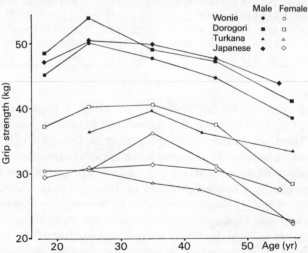

Fig. 1. Grip strength and aging in Wonie and Dorogori, in the Turkana (Little and Johnson, 1986), and in Japanese (Japanese Ministry of Education, Science and Culture, 1985).

Table 4. Correlation Coefficients between Maximum Grip Strength and Body Weight, Upper Arm Muscle Cross-Sectional Area (UMA), and Fat-Free Mass (FFM)

Group (N)	Body weight	UMA	FFM
Male			
Wonie (30)	0.535**	0.551**	0.565**
Dorogori (35)	0.635***	0.587***	0.722***
Total (65)	0.597***	0.582***	0.667***
Female			
Wonie (30)	0.406*	0.085	0.353
Dorogori (41)	0.320*	0.011	0.491**
Total (71)	0.529***	0.232	0.597***

* $p < 0.05$, ** $p < 0.01$, *** $p < 0.001$.

greater grip strength than Wonie males and females; this difference is particularly remarkable in females. Second, the grip strength of Gidra males (two village groups combined) is almost comparable with that of Japanese, and markedly (30–40%) greater than that of the Turkana. Third, Dorogori females' grip strength surpasses that of other female groups and even that of Turkana males. Fourth, Wonie female grip strength is comparable with the Japanese standard, with a higher value in the 30s and a lower value in the 50s.

It is recognized that muscle strength is related to muscle size (Malina, 1975; Little and Johnson, 1986), general muscularity (Clarke, 1966) and body weight (Clement, 1974). Thus, the correlations of Gidra grip strength with upper arm muscle cross-sectional area (UMA), fat-free mass (FFM), and body weight were calculated for sex/village groups (Table 4). The coefficients demonstrate that there is no significant correlation with UMA for women and that correlations with any variable are higher for men than for women. Judging from the correlation coefficients for two village groups combined, FFM has higher values than body weight does. Nevertheless, because there is not a significant correlation between FFM and grip strength for Wonie women and because FFM is not a measured but an estimated value, the present study treats the relationship between grip strength and body weight in order to examine inter-village differences. Figures 2 and 3 show scatter diagrams of males and females plotted for their grip strength and body weight, distinguishing them by village and by whether the value of BMI is greater than 24 or not. The figures show that the bulk of the individuals cluster around a straight line, but those with BMI of more than 24 tend to be located on the right side.

Regression analysis of grip strength versus body weight was done for all Wonie males, Wonie females, Dorogori males, and Dorogori females, and for Dorogori males and females whose BMI values did not exceed

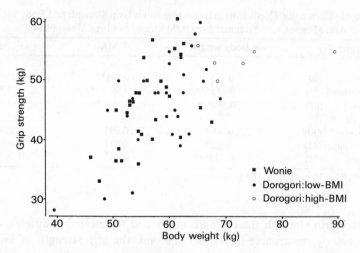

Fig. 2. Grip strength and body weight of Wonie and Dorogori males. The boundary between low and high body mass index (BMI) is 24.

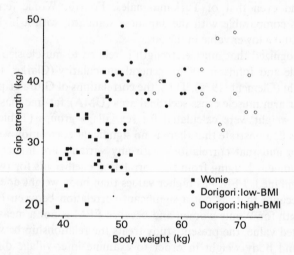

Fig. 3. Grip strength and body weight of Wonie and Dorogori females. The boundary between low and high body mass index (BMI) is 24.

24. The regression coefficients (r) for the three male groups range from 0.54 to 0.64 (p<0.01 in all cases), while those for the three female groups are low, ranging from 0.32 to 0.41 (p=0.059 for Dorogori women with BMI less than 24, and p<0.05 for the other two cases). The result is shown in Fig. 4. The inter-village difference in the slopes of regression lines for

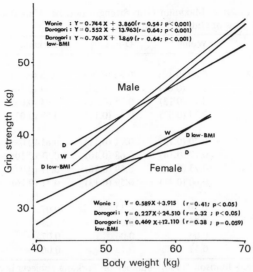

Fig. 4. Regression lines of relationships between grip strength and body weight for Wonie males and females and Dorogori males and females, and for Dorogori males and females with low BMI (a value of BMI less than 24).

either sex becomes smaller when high-BMI individuals are excluded, i.e. 0.744 for Wonie males versus 0.552 for all Dorogori males and 0.760 for low-BMI Dorogori males, and 0.585 for Wonie females versus 0.227 for all Dorogori females and 0.469 for low-BMI Dorogori females, although there is no statistically significant difference between the slopes.

DISCUSSION

A comparison of maximum grip strength among different populations may be valid when converted to grip strength per unit of body weight. Table 5 shows this indicator for ten-year age groups (from 20–29 to 50 +) by sex for Wonie, Dorogori, Turkana, and Japanese; as mentioned previously, the Turkana are a pastoral group and are characterized by lower male muscle strength in comparison with Caucasoid standards (Little and Johnson, 1986); and the Japanese standard can be viewed as an example of an industrialized population. With a few exceptions among women of the Gidra, grip strength per unit of body weight decreases with age.

From comparison of grip strength per unit of body weight, the inter-population difference of males is represented by [Gidra = Japanese > Turkana]; that of females, by [Gidra > Japanese = Turkana]. In light of

Table 5. Means (SDs) of Maximun Grip Strength (kg) per Unit Body Weight (kg) of Two Village Groups of Gidra, Turkana, and Japanese

	Wonie	Dorogori	Turkana[a]	Japanese[b]
Male				
20–29	0.89 (0.12)	0.84 (0.06)	0.68 (0.11)	0.79
30–39	0.82 (0.09)	0.76 (0.04)	0.69 (0.10)	0.77
40–49	0.79 (0.13)	0.75 (0.11)	0.64 (0.08)	0.74
50+[c]	0.73 (0.03)	0.71 (0.15)	0.59 (0.07)	0.71
Female				
20–29	0.70 (0.10)	0.71 (0.11)	0.61 (0.08)	0.60
30–39	0.77 (0.14)	0.68 (0.10)	0.61 (0.10)	0.60
40–49	0.73 (0.14)	0.69 (0.16)	0.57 (0.09)	0.57
50+[c]	0.60 (0.11)	0.59 (0.13)	0.51 (0.16)	0.53
Sex difference				
20–29	0.19	0.13	0.07	0.19
30–39	0.05	0.08	0.08	0.17
40–49	0.06	0.06	0.07	0.17
50+[c]	0.13	0.12	0.06	0.18

[a] From Little and Johnson (1986); number of Turkana subjects is respectively 67 and 84 for males and females.

[b] From Japanese Ministry of Education, Science and Culture (1985); number of Japanese subjects in each sex/age group exceeds 5000, and only means are mentioned.

[c] For Japanese, subjects aged 50–59 are treated.

Little and Johnson's (1986) remarks that among the Turkana, females' arm muscle strength has markedly developed while males' has not, one of the distinctive characteristics of the Gidra is greater grip strength in women. This is related to their working conditions; more than Turkana women, Gidra women are engaged in heavy work such as carrying harvested crops and firewood, pounding sago pith, and so forth.

Also noticeable is the age-related component in sex differences among the Gidra. In the 20–29 and 50+ age groups, the sex difference is great and fairly close to that in Japanese, while in the 30–39 and 40–49 age groups the sex difference is small and close to that in the Turkana. When we take into account age variation in anthropometric measurements (see Fig. 1 and Table 2), it can be judged that the marked sex difference in the 20–29 age group derives mainly from the men's relatively lean body composition, and the marked sex difference in the 50+ age group is mainly caused by the females' sharp decline in grip strength despite the small decrease in body weight.

Grip strength per unit of body weight is greater in Wonie villagers than in Dorogori villagers. The major reason for this is the greater body weight of Dorogori villagers, in particular of women. Our nutrient intake survey reveals that the energy intake is larger in Wonie than in Dorogori (Chapter

9 in this volume). Thus it can be considered that the greater body weight of Dorogori villagers is caused by lower energy expenditure (cf. Chapter 6 in this volume). Compared to Wonie villagers, Dorogori villagers less frequently practice energy-consuming hunting, sago-making, long-distance walking, and carrying of heavy things (Chapter 1 in this volume). The reduction in these activities is largely caused by modernization.

The relationship between grip strength and body weight becomes clearer from the regression analysis, although the regression coefficients for the female groups are not so high. The slopes of regression lines for Wonie villagers imply that an increase of 1 kg grip strength is gained by an increase of 1.34 kg and 1.67 kg body weight for males and females, respectively. The corresponding values for all Dorogori males and females are much greater: 1.81 kg and 4.41 kg. However, when only Dorogori villagers with BMI less than 24 are treated, these values decline to 1.32 kg and 2.13 kg, the former almost the same as that for Wonie males and the latter close to that for Wonie females.

It is recognized that the Wonie villagers have preserved the traditional way of living and the Dorogori villagers have been influenced by modernization. It can thus be concluded that in the modernization process grip strength per unit of body weight decreases and that this is largely attributable to the increase of body weight. Since grip strength per unit of body weight is very similar in Dorogori and Japanese males (Table 5), the relationship between grip strength and body weight among the Dorogori, at least in males, is judged to represent the pattern in modernized populations.

(*Ryutaro Ohtsuka*)

Chapter 4

Hunting Productivity

Since the epoch-making symposium "Man the Hunter" (Lee and De-Vore, 1968) stimulated research on the hunting-gathering way of life for the purpose of understanding human adaptation and evolution, a number of vigorous studies have been conducted among contemporary populations who depend fully or partly on hunting and gathering (e.g. Bicchieri, 1972; Lee and DeVore, 1976; Winterhalder and Smith, 1981). One of the aspects that still needs further information and analysis is detailed observation of hunting activities in local ecosystems, in particular the individual-based records of such factors as time spent in hunting, choice of hunting strategy, and hunting efficiency.

Based on the data collected in my field work in Wonie in 1971–72 and in 1981, this chapter focuses on the relationship between hunting activity and the individual hunter's age, sensorimotor functions, and behavioral characteristics.

HUNTING METHODS

The hunting methods in the area are primarily classified as those for individual hunting and communal hunting (for details, see Ohtsuka, 1983; Watanabe, 1975). Individual hunting is carried out throughout the year in both savanna and forest. Major game animals in savanna are the "grass wallaby" (*Wallabia agilis*) and two species of bandicoot (*Echimipera* spp.), and those in the forest and small patched woodland areas in savanna are the "bush wallabies" (*Dorcopsis veterum* and *Thylogale* sp.), pig (*Sus scrofa*), cassowary (*Casuarius casuarius*), and deer (*Cervus timorensis*); of these, the deer was newly introduced to the area from the west perhaps in the 1960s, and numbers have increased since the 1970s (Downs, 1972). The traditional hunting weapons are bow and arrows. Shotguns were introduced to the area around 1960; three married males in 1971–72 and four in 1981 owned shotguns. A shotgun is occasionally used by villagers other than the owner. The superiority of the shotgun to the bow and arrow is evident in hunting big animals like cassowary and deer and for birds perched in tall trees; among the Ye'kwana and Yanomamö Amerindians of the Upper Orinoco River, the superiority of the shotgun over the bow and arrow was observed in killing arboreal and volant animals (Hames, 1979). Based on a study among the Waorani in eastern Ecudador, however, Yost and Kelley (1983) reported that the greatest impact from the shotgun was seen in killing terrestrial quadrupeds. Because the Wonie villagers are not always able to purchase cartridges from the town, however, the use of shotguns is still limited. In hunts using a bow and arrow, a hunter is sometimes accompanied by dogs. Whether dogs are taken or not depends on the kind of animal being hunted. For example, a hunter takes dogs when he plans to kill pigs, whereas he never takes them when he plans to kill cassowaries by means of ambush. On the other hand, in individual hunts with shotguns dogs are not taken.

Communal hunting is practiced in the dry season when the savanna grasses wither. For a communal hunt, a party of from ten to 20 hunters is temporarily organized. Two to four members set fires at a number of points by burning fronds of wild palms, making a circle about 1–2 km in diameter. The rest of the members stand around the circle at intervals, and those setting the fires later join them. The hunters wait for animals running out of the circle. The principal game animals are grass wallabies

Table 1. Game Animals Killed during the Observation Periods[a]

		All methods		Communal		Bow-arrow		Shotgun	
Game	N	Mean weight	Total weight	N	Total weight	N	Total weight	N	Total weight
Cassowaries	4	50.3	201.0	0	0	1	46.0	3	155.0
Pigs	7	44.0	308.0	0	0	4	181.0	3	127.0
Deer	1	58.0	58.0	0	0	0	0	1	58.0
Grass wallabies	58	12.4	720.6	33	397.2	17	220.7	8	102.7
Bush wallabies	67	4.4	296.8	28	123.0	32	139.3	7	34.5
Bandicoots	26	1.8	45.5	18	31.6	8	13.9	0	0
Bats	12	0.3	4.1	7	2.1	5	2.0	0	0
Rat	1	0.8	0.8	1	0.8	0	0	0	0
Echidna	1	2.7	2.7	0	0	1	2.7	0	0
Middle-sized birds	10	1.7	16.6	0	0	3	5.7	7	10.9
Small-sized birds	16	0.3	5.5	0	0	16	5.5	0	0
Python snake	1	20.0	20.0	0	0	1	20.0	0	0
Monitors	3	7.4	22.2	3	22.2	0	0	0	0
Total			1701.8		576.9		636.8		488.1

[a] Thirty-eight days in 1971–72 and 42 days in 1981. Number, mean weight (kg), and total weight (kg) of the animals killed in these durations are given.

and bandicoots; bush wallabies and other forest dwellers are also killed when the burning circle contains patched woodland. In a communal hunt, bow and arrow are used exclusively and dogs are not taken. This study categorizes the hunting methods into three types: individual hunting with bow and arrow (bow-arrow hunting), individual hunting with shotgun (shotgun hunting), and communal hunting.

Table 1 presents the number and weight of each kind (one species or several species of similar kinds) of game animal killed during the two intensive survey periods, 38 days in 1971–72 and 42 days in 1981. In terms of number killed, bush wallabies are the largest group, followed by grass wallabies and then bandicoots; in terms of total weight, grass wallabies are the largest group, followed by pigs, bush wallabies, and cassowaries. Shotgun hunting tends to kill large-sized animals, communal hunting tends to kill small-sized ones, and bow-arrow hunting is in between.

MATERIALS AND METHODS

In the Gidra society, hunting is exclusively practiced by males. It is at the *kewalbuga* age-grade that a man starts to become an active hunter. Beyond the *kewalbuga* age-grade, there are three other age-grades: *rugajog*, *nanyuruga*, and *miid* (see frontispiece). Here, all the subject males are placed in one of four groups: unmarried (*kewalbuga*; 16–17 to late 20s in age), younger married (younger *rugajog*; late 20s to approximately 35), elder married (elder *rugajog*; 35 to late 40s), and elders (*nanyuruga* and *miid*; late 40s and over); they are called U, yM, eM, and E groups, respectively. This grouping based on the age-grade system corresponds

Table 2. The Subjects and Observation Time by Age Group

Age group	N	Observation time (hr) Mean±SD	Total
1971–72			
Elders (E)	5	242±188	1209
Elder married (eM)	6	323± 86	1940
Younger married (yM)	6	294±120	1764
Unmarried (U)	9	175± 83	1577
Total	26	250±133	6490
1981			
Elders (E)	5	311±184	1557
Elder married (eM)	9	433± 42	3899
Younger married (yM)	10	432± 93	4321
Unmarried (U)	8	425±152	3403
Total	32	412±127	13179

to that based on chronological age, with exceptions owing to the fact that a *kewalbuga* male sometimes marries earlier than his seniors. However, these exceptions are few, because the birth order of the people is well recognized and the traditional custom prescribes marriage according to birth order.

The subjects of this study numbered 26 in 1971–72 and 32 in 1981 (Table 2). Twenty of them were studied in both periods; three died during the intervening years and three others migrated out; three migrated in, one absentee returned, and eight younger males entered *kewalbuga* (two of them had married before 1981 to enter *rugajog*).

The quantitative data on hunting activities were collected as follows. In 1971–72, hunting activities were recorded during three time-recording surveys of all adult villagers' food-getting activities: for 12 days from 19 to 30 July, for 13 days from 11 to 23 October 1971, and for 13 days from 28 February to 11 March 1972. In these surveys, with aid of an assistant from Wonie village, I recorded times of departure from and return to the village from morning till evening for all adult villagers and asked these men upon their return where they had been and what they had done. In 1981, the same survey was done for 14 days from 18 to 31 October, and additional surveys were conducted for 28 days, i.e. from 25 September to 17 October and from 1 to 5 November; in them, I and my assistant did not observe the villagers' activities throughout the day, but asked them many times a day about their hunting practices.

To assess time allocated to hunting, this study defines observation time as the 14 hr between 6:00 and 20:00, from which time spent in overnight trips was substracted; for example, when a subject departed at 10:30 for another village to stay overnight the observation time for that day was treated as 4.5 hr. The observation time, broken down by age group, is shown in Table 2.

Most of the animals killed were carried to the village, although animals were sometimes dissected at the spots where they were killed. In all of the former cases and in some of the latter ones, the animals were weighed by me and/or my assistant. When we could not weigh the animal, we asked the hunter to estimate its weight in comparison with several animals of the same species that had been killed in the recent past.

Hunting grounds were identified by the hunters using local place names, more than 200 in number in the village-land. One named place was chosen as the hunting ground for each hunt, depending on the hunter's judgment of where he spent most of his time.

The sensorimotor functions of the hunters, which are analyzed in this paper, are visual acuity and grip strength. Visual acuity for each eye of the subject was examined using the international test chart of Landolt rings and was designated by a decimal notation that represents the reciprocal of the visual angle in minutes and is finally expressed in terms of Snellen's fraction given in meters: for example, 6/6 in meters = 1.0 or 6/60 = 0.1 (for details, Chapter 2 in this volume). For measurement of grip strength, a Smedley dynamometer was used to test the maximum voluntary contraction of the grip flexors (expressed in kilograms) of each hand of the subject in a standing position (for details, see Chapter 3 in this volume).

RESULTS

The number of observation days and the observation season differed between the 1971–72 and 1981 surveys. However, the hunting efficiency in terms of weight (in kilograms) of animals killed in the time (hours) spent hunting in both periods was almost identical for each hunting method: for bow-arrow hunting, 0.95 kg/hr in 1971–72 and 1.03 kg/hr in 1981; for shotgun hunting, 2.62 kg/hr in 1971–72 and 2.95 kg/hr in 1981; and for communal hunting, 0.97 kg/hr in 1971–72 and 0.61 kg/hr in 1981. Thus the data collected during the two periods are summed for the analyses of time spent in hunting and hunting efficiency in relation to hunting method and hunter's age group, whereas each hunter's records in the two periods are compared for analyses of change of hunting efficiency in ten years.

Table 3 shows hunting time per observation period, which covered 14 hr per day, from 6:00 to 20:00, broken down by method and by age group. When all methods are combined, hunting time per observation period was markedly low only in the E group. Broken down by method, a manifest inter-group difference in time spent in shotgun hunting is largely attributed to the ownership of shotguns and to the ability to purchase cartridges. More important facts are that bow-arrow hunting was practiced by the eM group more frequently than by the yM and U groups and

Table 3. Hunting Time (HT: hr) and Hunting Time per Daily Observation Period of 14 hr (HT/OT) by Method and by Age Group

Age group	N	Communal		Bow-arrow		Shotgun		Total	
		HT	HT/OT	HT	HT/OT	HT	HT/OT	HT	HT/OT
E	10	42.2	0.21	12.1	0.06	0	—	54.3	0.27
eM	15	189.5	0.46	259.7	0.62	100.8	0.24	550.0	1.32
yM	16	267.2	0.61	188.7	0.43	62.8	0.14	518.7	1.19
U	17	328.3	0.92	169.0	0.47	12.7	0.04	510.0	1.43
Total	58	827.2	0.59	629.5	0.45	176.3	0.13	1633.0	1.16

that communal hunting was practiced more frequently in decreasing order of age.

The weight (kilograms) of game animals killed per hour spent hunting, which is the direct indicator of hunting efficiency, is broken down by method and by age group (Table 4). The efficiency of shotgun hunting was 2.74 times that of bow-arrow hunting; this degree is fairly similar to that between shotgun hunting and bow-arrow hunting among the Ye'kwana and Yanomamö of the Upper Orinoco River, i.e. 2.31 times more efficient in the former (Hames, 1979). To compare the efficiency for the three methods combined among the age groups, this study calculated the corrected hunting efficiency to adjust the inter-group difference in time spent in higher productive shotgun hunting. In practice, based on the average efficiency of shotgun hunting, 2.77 kg/hr, and that of bow-arrow hunting, 1.01 kg/hr, the weight of animals killed in the former was multiplied by 1.01/2.77, or 0.37. The corrected values are also given in Table 4.

The interage-group difference in hunting efficiency for all methods, in terms of either game animals killed per hunting hour or its corrected value, is the highest in the eM group, followed by the yM group, and is fairly low in the E and U groups. Broken down by method, the efficiency of bow-arrow hunting is markedly higher in the eM group, being about twice that in the yM and E groups and five times that in the U group, whereas the difference in the efficiencies of communal and shotgun huntings are relatively small among the age groups, especially between the eM and yM groups (the efficiencies are a little higher in the yM group).

For the 20 hunters who were studied in both 1971–72 and 1981, each individual's hunting efficiency can be compared between the two periods,

Table 4. Game Animals Killed per Hour of Hunting Time (G/HT: kg/hr)[a] and Its Corrected Value (C-G/HT)[b] by Method and by Age Group

Age group	Communal		Bow-arrow		Shotgun		Total		
	N	G/HT	N	G/HT	N	G/HT	N	G/HT	C-G/HT
E	8	0.31	4	0.82	0	—	12	0.43	0.43
eM	33	0.80	88	1.60	44	2.74	165	1.54	1.22
yM	45	1.03	81	0.82	26	3.37	152	1.22	0.98
U	58	0.41	80	0.32	6	0	144	0.35	0.35
Total	144	0.70	253	1.01	76	2.77	473	1.02	0.84

[a] According to our analysis of composition of foods consumed by the Gidra (Chapter 7 in this volume), the mean energy and protein contents (per 100 g edible portion) of meats of 14 samples of major game animals are 97.7 kcal and 20.2 g. Taking the waste portion (40% on the average) into account, 1 kg of game animal provides approximately 590 kcal of energy and 12 g of protein.

[b] See the text for calculation of the corrected value.

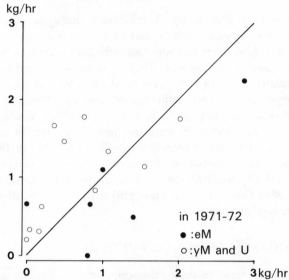

Fig. 1. The relation of the corrected value of animals killed per hour spent hunting in 1971–72 (on the abscissa) and that in 1981 (on the ordinate) for the 17 hunters.

although three who had belonged to the E group in 1971–72 killed no animal in the intensive study periods in 1971–72 and 1981. Figure 1 shows the scattergram of the corrected hunting efficiency in 1971–72 and that in 1981 for the 17 hunters (excluding the above three aged individuals). Of the 11 individuals who constituted the yM and U groups in 1971–72, eight increased in efficiency in 1981; whereas of the six individuals who constituted the eM group in 1971–72, four decreased in efficiency. Also noted is that each individual's efficiencies in the two periods correlate: when we calculate the rank correlation coefficient between the two efficiencies for the 17 individuals, Spearman's $r = 0.732$ ($p < 0.01$); in the same calculation treating the 20 hunters including the three aged individuals, Spearman's $r = 0.802$ ($p < 0.01$).

Each age group's mean grip strength of both hands and visual acuity (in terms of percentage of individuals with less than 0.8 in either eye) are shown in Table 5. The interage-group comparison demonstrates that grip strength was low in the E group and differed to a smaller extent among the remaining three groups and that visual acuity was also markedly low in the E group.

The hunters whose visual acuity in either eye was less than 0.8 numbered five in 1971–72 and also five in 1981. As shown in Table 5, all of them belonged to the E group except one individual who belonged to the eM group in 1981. During my observation periods only one of the five hunters in 1971–72 went hunting, and he killed no animal; three of the five hunters in 1981 went hunting, and only one of them (who belonged to the yM group) killed animals. These findings imply that visual acuity is closely related to aging and, as the result, also to time spent in hunting and hunting efficiency. To examine the relationship between grip strength and hunting efficiency among the hunters of the eM, yM, and U groups, the correlation coefficients between the mean grip strength and the corrected hunting efficiency of these hunters (21 in number in 1971–72 and 27 in 1981) were calculated: as the result, Pearson's correlation coefficient $r = 0.543$ ($p < 0.05$) for 1971–72 but $r = 0.166$ ($p > 0.05$, not significant) for 1981. Thus it is judged that grip strength and hunting efficiency are not strongly interrelated.

AGING AND SENSORIMOTOR FUNCTIONS

One of the concerns in the relationship between hunting activity and aging is the extent to which time spent in hunting differs among the age groups. As shown in Table 3, a notable characteristic is the low proportion of hunting time in the E group, about one-fifth that in the eM, yM, and U groups. Since there is no social restriction of the elders' hunting in the Gidra society, it is natural to consider their physical capabilities such as visual acuity and grip strength (Table 5) as the major reasons. As reported elsewhere, based on the records from four villages including Wonie (Chapter 2 in this volume), visual acuity among the Gidra, both men and women, is significantly low in the elder group, and the lowered acuity is correlated with the advance of cataracts (or corneal opacity). Grip strength among the Gidra, either its original value or its value per unit of body weight, begins to decrease in the 20s and becomes particularly low in the 50s (Chapter 3 in this volume), as in other populations (e.g. Clarke, 1966; Malina et al., 1982; Little and Johnson, 1986). These data suggest that

Table 5. Mean Grip Strength of Both Hands (mean ± SD) and Proportion of Individuals with Poor Visual Acuity[a] Classified by Age Group

Age group	N	Grip strength (kg)	Poor visual acuity (%)
E	10	34.2 ± 4.3[b]	90.0
eM	15	43.2 ± 5.7	6.7
yM	16	46.7 ± 5.1	0
U	17	43.0 ± 7.4	0

[a] Less than 0.8 in either eye.
[b] Significantly different from either of the other three groups at $p < 0.01$.

aged individuals' retarded visual acuity and grip strength are related to their short hunting time and lower hunting efficiency.

On the other hand, the younger boys who have not entered the *kewalbuga* age grade (the U group) sometimes go hunting in their own groups or, occasionally, together with the adolescents and adults, although their contribution to the food supply is negligible; this accords with the results of Kawabe's (1983) retrospective interview survey of boys and adolescents in a different Gidra-speaking village, Rual. The measurements of grip strength for the three oldest such boys in Wonie revealed that even the highest grip strength in terms of the mean of both hands was less than 30 kg. For this reason, even the strongest boy's bow was not as strong as an adult's.

Visual acuity and grip strength are directly related to the hunting practice of using bow and arrow: visual acuity is essential in finding animals or in recognizing, for instance, faint movements of grasses in which animals hide; grip strength is also essential in manipulating a bow, as mentioned above. Concurrently, however, it is possible to assume that these two represent sensorimotor functions or physical capabilities in a wide sense. It is thus concluded that sensorimotor functions limit the range of ages in which individuals are active and productive hunters—in the present case, between the late teens and about 45 years of age.

HUNTING STRATEGY AND BEHAVIORAL SKILL

One of the attractive results of this study was that although grip strength and visual acuity as well as proportion of hunting time were almost identical in the eM, yM, and U groups, hunting efficiency significantly differed among the three groups. For analysis of the interage-group difference, it is useful to assess the degree of difficulty of the hunting methods used. From a behavioral viewpoint, hunting is broken down here into three components: searching for animals, approaching the animals, and actually shooting an arrow or shotgun shells. Difficulty is arbitrarily categorized into three grades ($+$, $\pm$, $-$). Bow-arrow hunting is judged to be more difficult and to require more skill than the other two methods (Table

Table 6. Graded Difficulties in Hunting Performance Using Each Method

	Communal	Bow-arrow	Shotgun
Searching	$\pm$	$+$	$+$
Approaching	$-$	$+$	$\pm$
Shooting	$+$	$+$	$-$

$+$: Difficult, $\pm$: intermediate, $-$: easy.

6). It is thus reasonable that the interage-group difference in efficiency was the largest in bow-arrow hunting.

According to my observation and the hunters' judgments, two of the components, searching and approaching, determine to a large extent success or failure in hunting. The most important factors in searching for animals in individual hunts are 1) deciding on a hunting strategy in accordance with the season, the time of day, and the kind of animals to be hunted; and 2) awareness of minute and ever-changing environmental conditions and especially of the spots where animals are apt to hide. The approach to animals is particularly important in bow-arrow hunting. For instance, in hunting grass wallabies, because it is effective to shoot arrows from a distance of less than 20 m, and ideally from about 10 m, the hunter must approach the animal as close as possible from the lee side by walking slowly in a slouching posture and/or creeping. The importance of these two components is supported, if indirectly, by my finding for the same subjects, in the 1971–72 survey period, that the scores in a shooting test following the rules of the American Archery Round did not differ between unmarried males (the U group) and married males (the yM and eM groups combined), but the actual hunting efficiency was more than twice higher in the latter (Ohtsuka, 1977b, 1983).

As a behavioral factor that may relate to the interage-group difference in hunting efficiency, this study analyzed the use of space. In practice, the village-land was divided into meshworks of 1×1 km squares, with the village in the center, and one square was identified as the site of each

Table 7. Hunting Time (hr) per Daily Observation Period of 14 hr (HT/OT) and the Corrected Value for Game Animals Killed per Hunting Time (C-G/HT: kg/hr) by Hunting Ground Classified according to the Distance from the Village

Age group	HT/OT			C-G/HT		
	Near zone[a]	Remote zone[b]	Both	Near zone[a]	Remote zone[b]	Both
E	0.06 (23.0)	0.21 (77.0)	0.27	0	0.55	0.43
eM	0.54 (40.9)	0.78 (59.1)	1.32	1.00	1.37	1.22
yM	0.39 (32.3)	0.81 (67.6)	1.19	0.87	1.03	0.98
U	0.51 (35.6)	0.92 (64.6)	1.43	0.27	0.40	0.35
Total	0.42 (36.0)	0.74 (64.0)	1.16	0.72	0.90	0.84

[a] Near zone includes 9 km² or nine meshes of 1×1 km around the village.
[b] Remotezone includes other village-land.

hunt. Then the meshes were categorized into a "near zone" consisting of 9 km² around the village and the "remote zone" of the remaining area. The results are shown in Table 7. Hunting efficiency, indicated by the corrected value for the animals killed per hour of hunting time, was higher in the remote zone for any age group, by one-fifth on average; this is presumably because of the relative abundance of animals in the remote zone because of less disturbance by the people. In this connection, it is noticed that the proportion of time spent in the near zone was higher in the eM group than in the other groups. This implies that, in a minimum sense, the higher hunting efficiency of the eM group does not owe to space use differences with distance from the village.

The above discussions suggest that the higher hunting efficiency of the eM group is related to proficiencies in the two components, searching for animals and approaching them, and these behavioral proficiencies are achieved by a knowledge system expanded through experience. This parallels the judgement of Laughlin (1968), who emphasized the importance of knowledge of animal ecology and environmental conditions. It is thus concluded that the difference of hunting efficiency among the eM, yM, and U groups mainly comes not from their physical capabilities but from their behavioral skill associated with knowledge and experience, even though the reasoning remains to be concretely identified.

INDIVIDUALITY OF HUNTING EFFICIENCY

Besides the overall trend of hunting efficiency in accordance with aging, this study has revealed that hunting efficiency varies to a great extent among individuals. It must be noted that the present records covered only about 40 days in each study period; consequently the degree of individual variation itself will not be fully detected. Nonetheless, based on my experience in the study village for more than 15 months in total, it is safe to say that there is fairly large age-independent individual variation. This study has also revealed that the efficiencies of the 17 hunters in 1971–72 and in 1981 (all of whom belonged to the eM, yM, or U groups in both periods) are interrelated according to the rank correlation analysis. The individual hunters' characteristics are judged to play a decisive role in the individual variation.

The correlation analyses revealed that hunting efficiency was significantly correlated to the grip strength of the hunters belonging to the eM, yM, and U groups for 1971–72, but not significantly to that for 1981. Also noted is that almost all of the hunters who belonged to the eM, yM, and U groups had normal visual acuity. These results suggest that each hunter's abilities, which trigger the individual difference in hunting efficiency, depend to a lesser extent on physical strength. As mentioned previously, the searching for and approaching the animals, which tend to determine hunting efficiency, are apparently related to the hunter's behavioral skill more than to physical abilities. Thus, there is a possibility that, as pointed out by Watanabe (1971), intellectual faculty is a decisive factor in each individual's behavioral skill and thus plays a significant role in individual variations in hunting efficiency. Whatever the reasons are, this aspect is of great concern in understanding human adaptation and evolution; further studies should be devoted to the question whether large individual variation and the associated individualities are, in general, intrinsic parts of hunting efficiency.

(*Ryutaro Ohtsuka*)

Chapter 5

Productivity of Plant Foods

Throughout New Guinea, plants provide all populations with their basic foods. The staples comprise tubers (particularly, yam, taro, and sweet potato), *Metroxylon* sago, and banana, while supplementary plant foods, which include a number of wild plants, exceed 200 species (Powell, 1976). Aside from sago, the staples are species introduced to New Guinea, while the native plants enrich the secondary food register.

The sago plam is assumed to have spread originally from Indonesia into Melanesia (Barrau, 1959), and it may have colonized New Guinea a long time ago. Regarding the exploitation patterns of New Guinea tree crops as a whole, Yen (1974) pointed out that the connotation of "gathered" or "wild" is often applied to all, except for banana and coconut. Sago falls into an intermediate category; Golson (1977) has termed it a "minimally managed" resource in order to discriminate it from the polarized wild and cultivated categories.

The Gidra obtain their plant foods by various activities. If their plant-food-getting activities are, in general terms, grouped into the collecting of wild resources and the cultivation of domesticated resources, the utilization of sago and coconut combines both these aspects. Transplanting shoots of the sago palm is frequently practiced, as is that of coconut seedlings. Although these palm stands are sometimes visited by the owner, no special preparatory labor is required before the actual harvest, as is true in collecting wild plant food. Thus, plant food procurements can be divided into three groups: collection of wild plants, exploitation of sago starch and coconuts, and garden horticulture.

This chapter delineates major plant-food-getting activities and compares their productivities. For this purpose, we need observation records of human activities in both dry and wet seasons. Such records were collected only in an inland village, Wonie, during my 1971–72 survey period, and consequently the data of this chapter are exclusively based on the observation in this village. Despite the fact that subsistence patterns, as seen in time spent in various food-getting activities and in amount of foods consumed, differ among Gidra villages (e.g. Chapters 1 and 9 in this volume), each plant-food-getting activity is performed in a similar fashion in any village, and all the major types of activities throughout the Gidra territory are practiced by Wonie villagers.

Plant-food-getting activities treated in this chapter include collection of galip fruit (galip in Tok Pisin: *Canarium vitiens*) and that of tulip fruit (tulip in Tok Pisin: *Gnetum gnemon*), exploitation of sago, exploitation of coconut, and horticulture. Of the wild plant foods utilized in Wonie, galip and tulip fruits were largest in consumed amounts. Horticulture is not separately analyzed by crop since a variety of crops are usually planted in the same garden and the labor input for each kind of crop cannot be distinguished. The comparison is focused on the difference among the activity types; unlike hunting (Chapter 4 in this volume), marked difference of productivity among individual persons was not observed.

PLANT-FOOD-GETTING STRATEGIES

Galip Fruit

Galip trees grow naturally and are not cultivated. They are fairly abundant, and the villagers utilize those which are located within a half-hour's walk of the village. The tree is cut down with an iron axe by men, and women (and occasionally children) collect the fruits. The harvest season is restricted to the late dry season from September to November, and a collecting party will consist of one or two men, either unmarried (*kewalbuga*) or married (*rugajog*), and several women (and children) from different households.

Tulip Fruit

Tulip trees also grow naturally in the area, although it may be an introduced species and the trees are few in number. The harvest occurs in the early wet season from December to February. Since only fallen fruit is collected, gathering parties are rarely organized. It is usual for a woman, sometimes accompanied by her children, to collect the fruit for a short duration on such occasions as travelling from garden to village.

Sago

Sago palms grow in freshwater swamps, and the matured palm is 10 to 15 m in height and has a trunk 50 to 75 cm in diameter. The palm flowers and fruits only once during its life span, and continues to store starch in the trunk for a 10-to-15-yr period before flowering. After the tree bears fruit it gives out a number of shoots, and therefore sago stands tend to be dense under natural conditions. However, palms can also be reared through the transplanting of individual suckers. The harvest of the starch is done independently of the seasonal cycle of climatic conditions, and the palm must be cut down before it flowers, otherwise the starch degenerates. Comparisons of productivity between wild and planted sago stands among the Abelam (Lea,1964), the Sanio-Hiowe (Townsend, 1974), and the Gidra result in non-significant differences in the amounts of starch produced per unit of labor, but there is a considerable increase in the amount of extracted starch per tree in planted stands (Ohtsuka, 1977a, 1983).

Sago groves proliferate in several big, permanently inundated areas in Wonie territory, and most of the sago starch of the villagers comes from them. Naturally-grown stands and transplanted stands coexist in these groves. The transplanting is very easily done by individual men, usually while the women are engaged in starch-making. On the other hand, the extraction of the sago pith from the trunk is carried out by groups which comprise at least one male and one female, and usually include several members. It is the men's task to cut down the palm with an iron axe, and when it is lying on the ground to chop away the upper half of the bark, thereby exposing the pith which fills up the whole interior of the trunk. The women then begin scraping out the sago pith with a special wooden triangular-shaped tool, and pounding it into a fibrous mass. This work usually involves two or more women, each working on the same trunk at separate points. When a quantity of loosened pith is accumulated, the second stage of the operation, which involves washing out the sago powder from the pith, commences in a trough made out of a sago frond. The pounded pith is kneaded with water in the trough, and the suspended sago starch runs out with the water through a filter into a collecting vessel. The water can then be drained off.

The most notable feature of sago exploitation is that no care is necessary for the plant before the harvest, and the whole activity may be viewed as an aspect of collecting rather than horticultural behavior.

Coconut

Planting methods and the organization of harvesting parties for coconuts are almost identical to those for sago, and coconut palms also take from ten to 15 years to attain maturity. However, the coconut differs from the sago palm in two essentials. First, coconut palms grow on drier land, so it is much easier to expand the areas of coconut groves, although in fact only a small proportion of suitable land is exploited. Second, the harvesting of coconuts merely requires a man to climb the tree to obtain the fruits, which are then husked by women.

Horticulture

Sites under wooded forest or woodland savanna are selected for gardens, usually within a half-hour's walk of the village. Due to the seasonal growth patterns of some of the crops, particularly taro, horticulture requires a yearly cycle of labor. In the early dry season, members of each household clear one or two new large gardens, the men felling the trees with axes while the women clear shrubs and grasses with bush-knives. After the trees, shrubs, and grass have dried and been burnt, the gardens are occasionally fenced in to prevent damage by wild pigs, and planting then begins in the late dry season. Between planting and harvesting, weeding is done unsystematically. Harvesting requires little labor input and is conducted little by little, as the food is required. Horticultural labor is performed by men and women; fence building is done almost exclusively by men, while other activities are mainly done by women.

STABILITY OF PRODUCTION

Stability in the supply of food resources is essential in the hot and humid habitat of the Gidra, where food storage procedures are undeveloped. Sago starch and coconuts are available all the year round, whereas a large number of horticultural crops and all the wild fruits are harvested in particular seasons. Of the six major garden crops, taro, yam, cassava, and sweet potato can be harvested only in the dry season, whereas banana and papaya are harvested in both dry and wet seasons even though these non-seasonal crops are to some degree subject to seasonal fluctuations in productivity (Table 1).

Long-term availability also varies among plant foods, and major damage to them can result from abnormal climatic conditions, especially when

Table 1. Seasonal Availability of Major Plant Foods

	Dry season	Wet season
Sago	+	+
Coconut	+	+
Garden crops		
Taro	+	−
Banana	+	+
Yam	+	−
Papaya	+	+
Cassava	+	−
Sweet potato	+	−
Wild fruits		
Galip	+	−
Tulip	−	+

associated with damage by blight and noxious insects. Such natural calamities normally have greater effects on the annual vegetable plants like taro, yam, and sweet potato, than on the woody species like sago. On the other hand, as mentioned in Chapter 9 in this volume, coconut fruit is sometimes spoiled by the predation of giant rats (*Rattus rattus*) in the area. Comparing between sago and garden crops, particularly tubers, which are two major energy-providing plant food groups, the former far excells the latter in the availability.

LABOR PRODUCTIVITY

In this study, labor efficiency or labor productivity is assessed in terms of the amount of food energy involved in the products per unit of labor time, which has been used in the so-called input-output analysis developed by Carneiro (1957) and his followers. This measure possesses considerable validity in comparison between plant-food-getting activities, since plant foods act mainly as energy sources; another input-output measure, food energy in the products per energy expenditure of human labor (see Chapter 6 in this volume), will be analyzed in a forthcoming paper.

There is, however, a difficulty in this comparison because procurement of wild foods as well as sago and coconut requires little or no labor input other than harvesting, while horticulture involves making a new garden, planting, weeding, and harvesting, which continue in succession. Thus, two types of comparison have been undertaken in the present analysis. The first is a comparison among the productivity of sago exploitation, galip fruit collecting, and tulip fruit collecting (collecting of coconuts is not compared because of insufficient data). The input-output record of this comparison is based on my thorough observations in the working spots. Non-working time in the working places, for example, for resting and cooking, is excluded from the calculations.

The second comparison is made for identifying horticultural efficiency. For this purpose, horticulture and sago-making, both of which were frequently practiced, are compared for the time spent in all stages of activity observed during two 13-day time-recording surveys for all adult and adolescent villagers (see Chapter 1 for the methodological details) and the quantity of energy taken during two 12-day food consumption surveys for six households (see Chapter 9 for methodological details); both surveys were conducted once in the dry season, October and November, and once in the wet season, February and March (Ohtsuka, 1983). Taking sago exploitation as a link with horticulture on the one hand and with collection of wild fruits on the other, a synthetic comparison is possible, even if based on crude indications.

Table 2. Average Input-Output of Horticulture, Sago-Making, and Collecting of Galip and Tulip Fruits

	Horticulture	Sago[b]	Galip	Tulip
Calculation 1[a]				
Product (kg)/labor (hr)	—	1.9	2.9	4.6
Energy (kcal)/labor (hr)	—	3320	6550	4960
Calculation 2[a]				
Energy (kcal)/labor (hr)	980	1550	—	—

[a] *Calculation 1* is made with direct observation data, and *Calculation 2* is based on the data from time-recording and food-consumption surveys.

[b] In my previous report, on which this chapter is based, the values for energy (kcal) per labor hour of sago-making were higher because the energy content of sago starch was not derived from our own results (Chapter 7 in this volume) but from the literature.

Table 2 shows the result of two types of comparative calculations. Comparing the results of eight observations of sago-making with those of five observations of galip fruit collecting and of three observations of tulip fruit collecting (*Calculation 1* in the upper part of Table 2), the average energy return per man-hour of labor for the former is half or two-thirds of the latter. The lower productivity of sago-making comes largely from the complexity of technique, which includes pounding and washing out. Even if the labor efficiencies are recognized as identical, it can be concluded that the "elaborateness" of technology involved in sago-making does not make for higher productivity, in comparison with collecting of wild fruit.

The second comparison (*Calculation 2* in the lower part of Table 2) indicates that one man-hour of labor provides about 1000 kcal in horticulture and 1550 kcal in the sago exploitation. In general terms, it can be judged that the exploitation of sago is more productive than horticulture. Combining the above two comparisons, the three plant-food-getting activities are placed in decreasing order of labor productivity, as follows: wild fruit collection, sago exploitation, and then horticulture. In my view, this order parallels the increasing order of technological complexity and necessity of preparation before direct harvesting. Even if we ignore collection of wild fruits, which provide a smaller amount of food for Wonie villagers, the superiority of sago exploitation to horticulture is recognized as one of the important ecological factors in the subsistence adaptation of the Gidra population. Also emphasized is the coexistence of various kinds of activities whose productivities differ greatly. This means that the Gidra subsistence system depends largely upon how plant food resources can be exploited with less regards of productivity, as found in the pristine subsistence system in general.

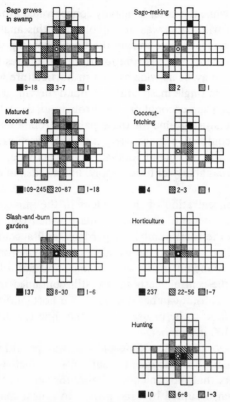

Fig. 1. Number of producing areas (on the left) and number of usages during two 13-day survey periods (on the right) for plant-food-getting activities on the meshed map of 1 km × 1 km, with the village in the center.

LAND PRODUCTIVITY

An area of 70–80 km² is recognized as the village-land of Wonie. In order to examine where the villagers exploit food resources, the village-land was divided into mesh squares of 1 km × 1 km with the village itself in the center. The analysis was applied to the distribution of sago groves (in swamp environments), coconut groves, and gardens in the village-land, and the frequencies of use for each mesh square were recorded during two 13-day time-recording survey periods, as mentioned above (Fig. 1). For these frequencies, the number of working visits to each mesh square was simply counted, without weighting for labor time or number of workers. The analysis discloses that the distribution pattern of sago groves and coconut groves differ greatly from that of gardens; the former is scattered widely while the latter is heavily concentrated around the village. The collecting of wild fruits was not involved in this analysis because of the small number of practices during the time-recording surveys. Judged from the scarce records, the distribution of mesh squares used in collecting of wild plants may be as wide as that in horticulture without marked concentration in the single mesh square around the village. The impartial distribution of sago and coconut groves throughout the village-land is related to the long maturing time for these palms. From a human ecological standpoint, the contrastive space-use patterns between sago exploitation and horticulture, the two major food-supplying activities, are considered in relation to the capability of food supply, like an idea of carrying capacity.

The spatial concentration of the gardens in the immediate vicinity of the village is very evident. The mesh square which contains the village includes 45% of the total number of gardens, and accounts for 65% of the usage. Woodland and densely wooded savanna environments suitable for gardens are abundant in the village-land, and according to my observations even the central mesh square still contained much unutilized land of this type. Thus, it can be concluded that horticulture requires only about 1 km² of land to be in use at any one time for the whole Wonie villagers, about 100 in number.

In contrast, sago groves luxuriate only in swamps, although the transplanting of suckers has expanded small less-productive groves along creeks. It appears that most niches suitable for sago in the village-land have already been occupied by sago groves to a near maximum degree, and it is reasonable to assume that these groves, which are scattered over an area of nearly 100 km², do not greatly exceed the minimum size required for a stable supply of sago starch to the Wonie population over a long-term period.

(*Ryutaro Ohtsuka*)

Chapter 6

Energy Expenditure

In 1947, Hipsley and Clements (1950) conducted a pioneer survey of food and nutrition in Papua New Guinea, and disclosed that per-capita daily energy and protein intakes of the sweet potato eaters in the highland were as low as 1600 kcal and 22 g, despite their well developed body physique. Then, based on energy costs of their major activities measured by means of indirect calorimetry and time allocation records, Hipsley and Kirk (1965) estimated their daily energy expenditure at 2100–2200 kcal for males and 1700–2000 kcal for females, a little higher than the intake level. The discrepancy between the energy intake and expenditure may be, to a large extent, attributable to overestimation of the energy costs for some kinds of activities, which were not measured but were represented by those obtained in other populations. This suggests that energy cost of the "same" activity, e.g. digging the ground with stick in horticulture, varies from population to population perhaps because of the difference in actual procedure. In this connection, it is noted that environmental conditions in Papua New Guinea, such as the altitude and the degree of undulations, greatly differ. Also important is the regional variation in the major subsistence activities: cultivation of sweet potato in the highland, taro or banana cultivation in the foothill area, and sago gathering in the lowland.

It is thus meaningful to investigate energy expenditure among the Gidra, and we have had special interest in their energy expenditure in relation to our previous findings. In our surveys in the Gidra, the energy intake and the time allocation in daily activities were investigated in 1971–72 and in 1981–82. One of the striking features was high energy intake, from 3000 to 3500 kcal per day per adult male among the four villages (Chapter 9 in this volume; see also Hyndman et al., 1989). On the other hand, the energy costs for some of the Gidra people's major activities, sago-making and group hunting in particular (e.g. Chapters 4 and 5 in this volume), have not been investigated in other populations.

In 1986, I did an energy expenditure survey among the Gidra. For this survey, I applied the heart rate method, theoretically based on the linear relationship between heart rate and energy expenditure, which has widely been used in field surveys (Bradfield *et al.*, 1969), with minor revisions for the inconvenient field conditions among the Gidra. When heart rate is continuously monitored, not only energy cost of each activity but also 24-hr energy expenditure can be calculated. The estimation error of this method was revealed to be compatible to other methods applicable to the activities outside the laboratory, about 20% for an individual person's daily energy expenditure. And the average estimation error for the population as a whole ranged within a few percent of the referred values obtained by an accurate laboratory method (Kalkwarf *et al.*, 1989).

SUBJECTS AND METHODS

The present data were collected in the period from September to December 1986, when I visited four villages, Dorogori, Ume, Wonie, and Rual, to stay for one to three weeks in each of them. First, I explained to the village people about the aim of this study and demonstrated how to use the heart rate recording machine (Memory Mac, Vein Co. Ltd., Japan). Seventy-five adults or adolescents in the four villages, more than half of the total number excluding children, voluntarily participated in this study. The subjects consisted of four *nanyuruga*, 15 *rugajog*, and four *kewalbuga* for males, and two *nanyukonga*, 11 *kongajog*, and three *ngamugaibuga* for females (see frontispiece for the Gidra age-grades), and both male and female subjects came from the four villages except females from Rual.

The heart rate of each subject was recorded once for a 24-hr duration from evening to the next evening, or for a 12-hr duration from morning to evening. In this setting, the subjects were instructed to behave as freely as they intended; however, a few villagers who had intended not to work in the proposed duration did not participate in the study. Throughout the study duration, the heart rate recording machine, about 180 g in weight, was attached to the waist of the subject with a belt and three electric codes extended from it were fixed in different parts of the chest and abdomen. The monitored heart rates were automatically sent to a hand-held computer (HC-40, Epson Co. Ltd., Japan) through an interface (Mac Reader, Vein Co. Ltd., Japan) and were stored in a micro-cassette tape. Simultaneously, for 39 out of the 75 subjects their activities were thoroughly observed by me at least during the day time, and when his/her activity changed its time was recorded by minute. From the observation records, per-minute heart rates in the monitored records were allocated to activity types, and then energy cost of each activity was determined.

When 24 or 12 hrs passed in the study, each subject was requested to perform a standard step test, using a step of 30–45 cm in height (the height differed from village to village because of the available step). The subjects were instructed to repeat stepping up and down at a constant speed of 15–20 steps per min (the instructed speed differed according to age of the subject). The subject person performed this test for 3 min, and the heartbeat during the last 1 min (when it was stabilized) was recorded.

For the data analysis, two heart rate values for each individual, i.e., one during resting time (except the times in which his/her heart rate was elevated by the preceding activities) in the monitoring survey and the other during the step test, were plotted agaist the respective energy expenditure values: the energy expenditure at resting time which was reported in the work of Norgan *et al.* (1974), and the energy expenditure estimated from the load and body weight during the step test in this study. Using this regression line, each individual's energy cost for each activity and his/her 12- or 24-hr energy expenditure were determined from the respective mean heart rates. The methodological assessment of this procedure is discussed in detail in another article (Inaoka, in press).

ENERGY COST OF EACH ACTIVITY

Each subject's activities recorded in the observation survey were grouped into categories, and the mean value of heart rates for each activity category was determined; in other words, when the activities included in the same category were observed more than once during the 12 or 24 hr, the mean heartbeats was calculated from the whole number of per-minute heart rates for the activities concerned. The total number of such activity categories was 46. Of these, 17 categories, which comprise time-consuming components of five major subsistence activities for the Gidra, are shown in Table 1, with the mean energy costs; the energy costs of the other activity categories are reported elsewhere (Inaoka, in press). The corresponding values in two previous studies in Papua New Guinea (Hispley and Kirk, 1965; Norgan *et al.*, 1974) are also mentioned in Table 1 for comparison.

Compared with the previous studies, the energy costs in this study were generally high, although those in three activity categories for coconut utilization were fairly identical. Because of the lack of detailed description on the activities in the previous reports, the reasons for the differences are difficult to discuss. However, there is a possibility that the subjects of the present study may have behaved more freely than those in other studies since the recording apparatus was light; in this connection, the similar level of energy costs for coconut-utilizing activities can be ex-

Table 1. Mean Energy Costs of Major Activities (cal/kg·min)

Category[a]	Present data[b]	Reference-1[c]	Reference-2[d]
Communal hunting (M)	105 (3)		
Individual hunting (M)	68 (1)		59–63
Sago-making			
Cut sago tree (M)	131 (4)		
Scrape out skin (M)	118 (4)		
Cut sago frond (M)	108 (7)		
Pound sago pith (F)	124 (4)		
Wash out starch (F)	104 (4)		
Coconut-fetching			
Collect fruit (M)	82 (2)		80
Husk skin (M)	100 (3)		108
Squeeze fruit (F)	67 (1)		69–75
Horticulture			
Cut sticks (M)	88 (2)		73–77
Strike stick (M)	116 (6)		
Thrust fence post (M)	93 (4)		80
Tie fence post (M)	87 (4)		56–57
Dig with hoe (F)	102 (5)	72–110	
Dig tuber (F)	86 (6)	48	53–54
Plant tuber (F)	75 (5)	50	64–83

[a] (M): male's activity, and (F): female's activity.
[b] Number of persons observed is shown in parentheses.
[c] Hipsley and Kirk (1965).
[d] Norgan et al. (1974).

plained because these activities normally continue for a short duration and should be done with similar tempo and speed in any groups, irrespective of the kinds of recording apparatus attached.

The energy costs for group hunting and various categories of sago utilization cannot be compared with other data. According to our observation, the higher level of energy cost in communal hunting than in individual hunting came from the different behavioral pattern. In group hunts, the hunters frequently ran to chase animals which came out of the firing area (see Chapter 4 for description of hunting activity), whereas in individual hunts walking was predominant. Any activity categories involved in sago-making were highly energy-consuming; the behaviors in them, in terms of movement of body muscles, were similar to those in men's activity of "strike sticks (into ground)" in horticulture and a women's "dig (ground) with hoe." In fact, energy costs in any sago-making activity were almost identical to or higher than the above-mentioned horticultural activities.

In actual setting, different components in each activity (for instance, "pound sago pith," "wash out (sago) starch" and some other activities (including even short-term pauses in women's sago-making) intermittently

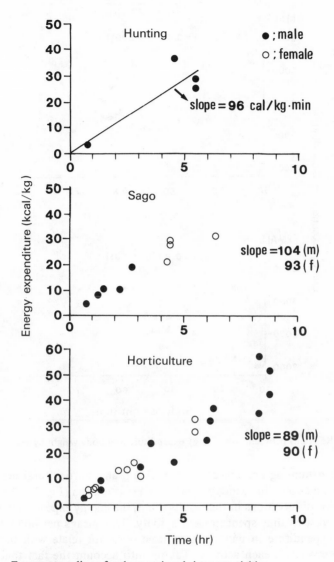

Fig. 1. Energy expenditure for three main subsistence activities.

occur. Thus, it is useful to estimate the mean energy expenditure for the whole activities, or for the whole duration of a set of activities, at the major category level. Based on each individual's records, the mean energy expenditures per 1 kg body weight for hunting (communal hunting and

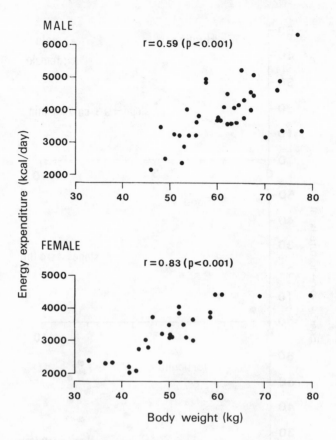

Fig. 2.   Relations between 24-hr energy expenditure and body weight by sex.

individual hunting combined), sago-making, and horticultural activities and the time spent in them are plotted in a scattergram (Fig. 1). There are two major observations in the results. First, the energy expenditure is well correlated with time spent in each activity. This means per-body weight energy expenditure in unit of labor time does not relate with the time length devoted to each activity. Taking into account the fact that each plot comes from different subjects, this high correlation suggests that when the time spent in each activity (in the major category level) is known the energy expenditure can be well estimated. Second, the sexual difference in energy expenditure in sago-making and in horticulture is small, despite the fact that the contents of activities are different. This may be one of the basic characteristics in the Gidra activity system, and is inconsistent with the general statement that the women's work is less energy-consuming. Particularly when the longer labor time of women in sago-making is taken into account, their high energy expenditure should be important in considering the adaptive mechanisms of populations like the Gidra who heavily depend on sago.

TWENTY-FOUR-HOUR ENERGY EXPENDITURE

For the estimation of 24-hr energy expenditure, one way is to calculate it from energy cost and precise time spent in each activity even in the major category level. However, the present analysis treats 24-hr energy expenditure of the Gidra subjects, which could directly be derived from the mean value of per-minute heart rates monitored for 24 hr; for the subjects whose heart rates were monitored for 12 hr (in the daytime), the per-minute heart rate in the other 12 hr was replaced with that in "resting" time, as mentioned previously. Thus, in this analysis the subjects whose activities were not observed by me were involved. The calculated daily energy expenditure was significantly correlated with body weight for either sex (Fig. 2).

The results demonstrate that daily energy expenditure for either sex differed by more than two times among individual adults or adolescents, so that the per-person level not only in energy expenditure but also in energy intake should be assessed, taking their body weight into account. Also noticeable is the fact that the mean daily energy expenditure was about 3900 kcal for males and about 3200 kcal for females. The values are fairly high compared to other findings, but very comparable with the Gidra energy intake levels (according to our records in 1981–82, the energy intake per day per adult male having the mean body weight of 55.7 kg was about 3000–3500 kcal among the four villages, as shown in Chapter 9 of this volume).

(Tsukasa Inaoka)

II. THE ECOLOGY OF FOOD CONSUMPTION

Overview

In understanding human nutrition, ecologists are concerned with a number of things which take place prior to digestion and absorption of nutrients through foods. The kinds and amounts of foods consumed and the amounts of nutrients they contain are the relevant factors, together with such information as what foodstuffs are available and how they are exploited in the local habitats. In the South Pacific region in general, where cereals, legumes, and milking animals were not known prior to contact with the West and still are only rarely consumed, a large number of plants and animals have been involved in the food register; however, basic knowledge about the amount of each food consumed by the local population and the composition of various local foods in their diet has been lacking. Thus, one of our objectives was the sampling and chemical analysis of the Gidra foods. The results for major nutrient composition (Chapter 7) reveal several analytical and ecological concerns, to which less attention has been paid. Chapter 8 emphasizes the interrelatendness of major nutrients and elements (micronutrients) in the foods, paying attention to the ecosystem framework. Due to methodological restrictions, however, the contents of vitamins in these foods have not been determined.

Using the results of food composition analyses, we estimated the nutrient intake of the people on the basis of the records of food consumption surveys in four villages in 1981 and in one village in 1971. Inter-village differences in intake of major nutrients (Chapter 9) and of micronutrients (Chapter 10) are related to local environmental conditions, subsistence patterns, and the modernization process. In our research framework, major and micronutrient intakes are regarded as the most basic information for assessing the people's biological and health states, as is shown in Chapters 11 to 17 under the heading "Nutrition and Health."

Chapter 7

Major Nutrients in Foods

Nutrient intake is, in general, estimated by applying food composition tables to the amount of foods consumed by the subject people. However, such food composition tables do not include some of the foods which are consumed by less modernized populations like the Gidra, and the reported values sometimes differ from the true ones because, for instance, of the conditions of the food specimens sampled. In this chapter, energy and major nutrient composition of foods consumed by the Gidra people are examined. Micronutrients or elements contained in foods are analyzed in the following chapter (Chapter 8), in connection with major nutrients. The foods which were sampled and whose composition was analyzed numbered 78. Of these, 70 were local products and eight were purchased foods which have gradually been added to the menu since several decades earlier. In selecting the foodstuffs for analysis, we had two different intentions: to collect as many staple foods as possible, and to collect foods whose composition had not yet appeared in the literature, even if the amount consumed was small.

METHODOLOGY

Sampling
Our research team members collected food samples during the survey period in 1980 and in 1981–82. The sampling of local foods was done in four villages, Rual in the north (by T.K.), Wonie inland (by myself), Ume in the riverine zone (by T.A.), and Dorogori on the coast (by T.I.); samples of purchased foods were obtained in a town market at the end of the 1981–82 survey period.

The local foods were collected in the villages in the state in which they were usually eaten. The collected samples were immediately weighed with a beam balance (measuring quantities as small as 0.1 g), and, when necessary, inedible portions were removed and separately weighed. Then the samples were cut into small pieces and dried in sunlight for periods ranging from several days to a couple of weeks so that they could be trans-

71

ported to the laboratory in Japan. Before packing, the dried samples were weighed again. The weight of each sample in the dried state was designed so as to exceed 50 g; three samples, however, were much lighter due to the scarcity of the foodstuffs, so that they were analyzed only for elements. It should be noted that the need for sun-drying made it impossible to collect succulent foods such as papaya, pineapple, and coconut, which were fairly frequently consumed in the area (in our 1986 survey period, some of these foods were sampled for analysis of element concentration, as mentioned in Chapter 10 in this volume).

After being transported to Japan, each sample (except for the three small samples) was divided into two portions, one for analysis of the composition of major nutrients (such as moisture, protein, carbohydrate, fat, fiber, and ash) and another for analysis of the concentration of elements.

Analysis
The chemical procedures of the analyses were as follows:
Moisture.   An accurately weighed sample (ca. 2 g) was heated at 135± 2°C for 2 hr in an electric oven with air circulation until there was no further loss of weight.
Protein (Nitrogen).   Total nitrogen was determined by Kjeldahl's method and converted to protein using the factor 6.25. For some samples, mainly of animal origin, the protein content was determined by the method of Barnstein (1900) to check the values estimated from nitrogen content.
Fat.   Crude fat was extracted by anhydrous ether in a Soxlet apparatus.
Ash.   Ash content was determined by ignition in an electric furnace at 550°C for 24 hr
Fiber.   Crude fiber was determined according to the A.O.A.C. method (A.O.A.C., 1970).
Carbohydrate.   Non-fibrous carbohydrate was calculated by substracting the sum of the amounts (g) of moisture, protein, fat, fiber, and ash from 100 g. Theoretically, the direct measurement of various carbohydrates would make possible fuller determination of food composition, but our analysis followed the "difference method" which has been used for a number of international and national food composition tables (for example, for African foods, FAO/USDHEW, 1968; for East Asian foods, FAO/USDHEW, 1972; for Japanese foods, Resources Council, 1982).
Food Energy.   Food energy was computed from a value of 4 kcal/g (16.7 kJ/g) protein, 4 kcal/g (16.7 kJ/g) carbohydrate, and 9 kcal/g (37.7 kJ/g) fat.

RESULTS

The nutrient composition per 100 g edible portion of the foods is shown in the Appendix. Regarding the composition of major nutrients, especially energy and protein, the following discussion will point out some analytical and ecological problems.

WATER CONTENT OF FOODS

In comparison with several published food composition tables, including those for the Pacific peoples (Hodges *et al.*, 1950; Peters, 1957; McCance and Widdowson, 1960; Bailey, 1968; FAO/USDHEW, 1972), the water content of our samples of several kinds of foods, especially sago flour, was high. As pointed out by Oomen (1971), the water content of raw sago varied from 20% to 45%, and this may depend on the time elapsed since production. According to my measurement, a 25% weight loss of sago flour was observed over the duration of a 30-day drying process, so that the water content changed from 44.5% on the day of production (to be precise, several hours after the production) to 26.3% on the final date (Table 1). This range is similar to the data of Oomen (1971). In contrast to our samples, most sago samples whose composition is mentioned in the literature might have been obtained in markets or shops, since their water content ranges between 12.5% and 20%. Thus, in applying the food composition of sago flour to the calculation of nutrient intake among local populations like the Gidra, the changing water content must be taken into account.

The water content in our samples of sweet potato and tree leaves of *Gnetum gnemon* (jointfir) is also greater than that in the literature, although the difference is only about 10%. The importance of water content of food was stressed by Hudson *et al.* (1980), who analyzed the diets of Gambian people and found a highly significantly negative correlation between water content and energy or protein content of various types of foods.

Table 1. Changes of Water Content in Sago Flour during Drying. Both rows are average values of two specimens.

No. of days from production	0	5	10	15	20	25	30
Weight (%)[a]	100	88.3	83.8	81.3	79.3	76.0	75.3
Water content (g)[b]	44.5	37.1	33.8	31.7	30.0	27.0	26.3

[a]Weight (as percentage of the initial weight), [b]Water content (g) per 100g.

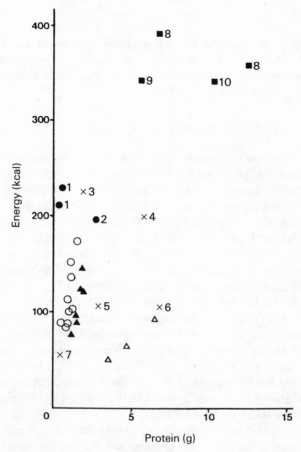

Fig. 1. Protein content and energy content of plant foods.
●: Sago (1: *Metroxylon* sago, 2: *Areca* sago), ▲: Tubers, ○: Bananas, ✕:
Fruits and nuts (3: meat of *Canarium* fruit, 4: cycad nut, 5: meat of *Gnetum*
fruit, 6: lotus seed, 7: mango), △: Leaves, ■: Purchased foods (8: two kinds
of biscuit, 9: rice, 10: wheat flour).

SIMILARITIES AND DISSIMILARITIES AMONG LOCAL FOODS

The bulk of the food intake of the people of Gidra consists of local
foodstuffs. By tradition, the major local foods are classified into several
groups in accordance with subsistence strategies, as follows (see Ohtsuka,
1983: 107): wild plants; sago; coconut; tuberous crops (e.g. taro, yam,
elephant-foot yam, and sweet potato) and banana; papaya and pineapple; land animals; and fish and other aquatics. While coconut, papaya,

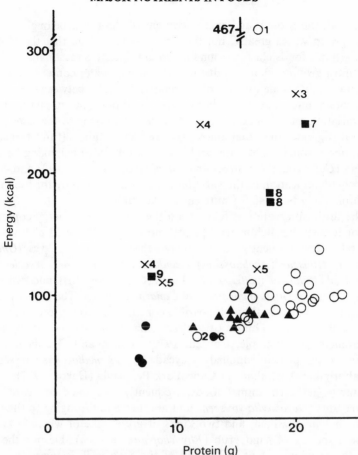

Fig. 2. Protein content and energy content of animal foods.
○: Land animals (1: fat layer of cassowary, 2: water snake), ▲: Fish, ●: Shellfish and prawn (6: prawn), ✕: Insects (3: unidentified grub, 4: sago grub, 5: two species of tree ants), ■: Purchased foods (7: corned beef, 8: two kinds of canned mackerel, 9: canned cream).

and pineapple were not sampled in our study, their consumed amounts were much smaller than those of the foods in the other categories.

Figures 1 and 2 show scatter diagrams of the plant foods and animal foods plotted for their protein and energy content. Three observations can be made on the basis of these diagrams. First, distinguishing the foods by protein and energy content results in considerable correspondence with the traditional Gidra classification. Special attention should be paid to the similarity between bananas and tubers.

Second, the protein and energy content of the foods belonging to the same group varies greatly, but the figures demonstrate that the plots of each group's foods cluster around a straight line. It is reasonable to judge that intra-group variation is due to difference in water content, and inter-group variation is mainly due to difference in the ratio between the amount of protein and that of carbohydrate. It should be emphasized that the variation among nine samples of banana belonging to the same genus (*Musa*) is greater than that among six samples of tubers (three taros, one yam, one elephant-foot yam, and one sweet potato) belonging to four genera (*Colocasia, Dioscorea, Amorphophallus*, and *Ipomoea*), and consequently identification of the varieties is important not only for food composition study but also for nutrient intake study.

The third observation is focused on the protein and energy content of minor foodstuffs, such as wild plants and insects. The wild plants which are relatively frequently eaten involve three kinds of leaves (*Gnetum gnemon, Hibiscus (Abelmoschus) manihot*, and *Ormocarpum orientale*), lotus seed (*Nelumbo nucifera*), cycad seed (*Cycas circinalis*), and two kinds of wild fruits of *Gnetum gnemon* and *Canarium vitiense*. The higher protein content in the leaves shows a possible contribution to the protein supply for the Papua New Guinean peoples, as has been pointed out by many researchers (for example, Corden, 1970; Oomen and Grubben, 1978). Similarly, the protein supplied by cycads (*Cycas media*) was stressed for the aboriginal Australians in Cape York Peninsula (Harris, 1977).

Among gathered animal foods, frequently consumed are two kinds of tree ants (*Oecophylla virescens* and another one belonging to the same genus but unidentified) and two kinds of grubs, one of which is known to be a species of sago grub (*Rhynchophorus schach*). Despite the fact that the protein content per weight of the sago grub is lower than that of game animals, the importance of the sago grub as a protein source has been emphasized in some of the literature on the Papua New Guinean peoples, especially in Fountain's (1966) report on a Sepik population. The fat-supplying role of *Rhynchophorus phoenicis* for the Angolan people was emphasized by Oliveira *et al.* (1976), who also pointed out its considerable amount of zinc, although the zinc concentration of the Angolan *Rhynchophorus* is about ten times that of the variety consumed by the Gidra (cf. Chapter 8 in this volume).

LOCAL FOODS VERSUS PURCHASED FOODS

Of the foods purchased by the Gidra, most frequently eaten were rice, (wheat) flour, canned mackerel, and corned beef. Rice and flour were substitutes for energy-supplying plant staples, especially sago flour, while

Table 2. Protein (g) per 1000 kcal Energy among Plant Staples

	N	Protein (g) per 1000 kcal energy
Sago 1	1	2.2
Sago 2	1	1.4
Tubers	6	14.6±1.4
Bananas	9	8.8±1.7
Rice	1	16.3
Flour	1	30.1

canned mackerel and corned beef were usually eaten together with rice or flour.

The energy content of rice and flour is higher than that of the local plant staples. The amount of protein (g) per energy unit (kcal) of the plant staples are shown in Table 2. According to this ratio, the protein supplied by flour is twice that supplied by tubers when the same amount of energy is taken. Practically speaking, more important is the fact that flour and rice have substituted mostly for sago flour, which contains a negligible amount of protein. The Gidra people, particularly the inlanders, have been basically sago-eaters; according to our food consumption study (Chapter 9 in this volume), half or more of food energy came from sago in Rual (northern) and Wonie (inland) villages. The poor protein content of sago is recognized as a vulnerable factor in the sago-eater's adaptation (Ohtsuka, 1977a, 1983; also see Oomen, 1971; Townsend, 1974; Ulijaszek, 1982).

The protein content of canned mackerel and corned beef is similar to that of local fish and land animals, while the fat content is much higher, and this increases the energy content.

The addition of these purchased plant and animal foods to the Gidra diet has been changing their nutrient intake and thus their nutritional status. As will be noted in our discussions of food and nutrient intake (Chapter 9 in this volume) and nutritional status (e.g. Chapters 11, 14, and 15), the changing food consumption pattern is recognized as one of the most important ecological conditions derived from modernization of the Gidra (cf. Dennett and Connell, 1988).

(Ryutaro Ohtsuka)

Chapter 8

Micronutrients in Foods

It has been reported that regional human nutritional abnormalities are related to the trace element levels in locally produced foods (e.g. Allaway, 1986; Crounse *et al.*, 1983a, 1983b). In the regional ecosystem, enrichment of elements occurs through biogeochemical circulation from soil through plants to animals and finally to humans. It is reasonable to consider that in this process relative abundances of trace elements differ from group to group in plants as well as in animals. Thus, in studying the nutritional status of a human population, it is necessary to know the compositional character by different food category in terms of trace elements as well as major nutrients.

The major nutrient composition (moisture, energy, protein, fat, carbohydrate, fiber, and ash) in 78 foods consumed by the Gidra was reported in Chapter 7 in this volume. In addition, we measured 17 elements, i.e. sodium (Na), magnesium (Mg), aluminum (Al), phosphorus (P), potassium (K), calcium (Ca), vanadium (V), chromium (Cr), manganese (Mn), iron (Fe), nickel (Ni), copper (Cu), zinc (Zn), strontium (Sr), cadmium (Cd), mercury (Hg), and lead (Pb) concentrations in the same food samples (see Appendix). Based on 17 elements and major nutrients, the present chapter aims to clarify the compositional character of the foods using a principal components analysis, one of the multivariate analyses which are used to reveal the pattern of associations between many variables. A similar analysis was applied to the interrelationship of elemental concentrations in shellfish (Favretto and Favretto, 1984b).

METHODS

Sodium (Na), Mg, Al, P, K, Ca, V, Mn, Fe, Ni, Cu, Zn, Sr, and Cd concentrations were measured by inductively coupled plasma atomic emission spectrometry, and Cr and Pb concentrations were measured by graphite furnace atomic absorption spectrometry after digestion by nitric acid in a Teflon-lined, high-pressure decomposition vessel (Uniseal Decomposition Vessels, Ltd.); mercury (Hg) was measured using the modified Magos method (Magos, 1971; Yamamoto *et al.*, 1980). Accuracy of the determination was checked by measuring Bovine Liver (NBS 1577). Elemental concentrations were calculated on a wet weight basis, on a dry weight basis, and as nutrient density per 1000 kcal; the values for wet weight basis are illustrated in this chapter.

Principal components analysis with varimax rotation was conducted using the SPSS-X program package (SPSS-X™ User's Guide, 1988). Elemental concentrations on a wet weight basis, on a dry weight basis, and as nutrient density per 1000 kcal were converted to logarithmic values for analysis.

MOISTURE, ENERGY, AND MAJOR NUTRIENT CONTENT

Moisture content and energy and major nutrient (protein, fat, and carbohydrate) content on a wet weight basis (data are recapitulated from Chapter 7 in this volume) by food category are shown in Fig. 1. Foods are classified into 12 categories: mammals (including corned beef), birds, reptiles, fishes (including canned mackerel), shellfishes, crustaceans, insects, starchy foods (sago flours, tubers, rice, wheat flour, and biscuits), seeds and nuts, leaves, fruits, and other foods.

Animal foods contained more protein and less carbohydrate than plant foods. Among animal foods, fat and energy contents in insects were high.

Moisture contents were extremely low in purchased starchy foods (rice, wheat flour, and biscuits). Corned beef and canned mackerel were also lower in moisture than local mammal and fish foods. Among the starchy foods, energy, protein, fat, and carbohydrate contents were higher in purchased foods than in local foods. Fat and energy contents of corned beef and canned mackerel were also higher than those in local animal foods.

TRACE ELEMENT CONCENTRATIONS

Concentrations of 14 elements (Na, Mg, Al, P, K, Ca, Cr, Mn, Fe, Cu, Zn, Sr, Hg, and Pb) on a wet weight basis by food category are shown in Fig. 2. Because they were detected only in three to seven samples (leaves, shellfish, and nuts; see Appendix), V, Ni, and Cd are not shown in the figure, nor are they included in the principal components analysis.

Five elements (Na, Mg, P, K, and Ca) shown in Fig. 2a were found at relatively high levels in foods. Sodium and P were contained at higher levels in animal foods than in plant foods. Purchased foods (except rice) were rich in Na. Among the starchy foods, P concentrations were higher in purchased foods than in local foods. Magnesium and K concentrations

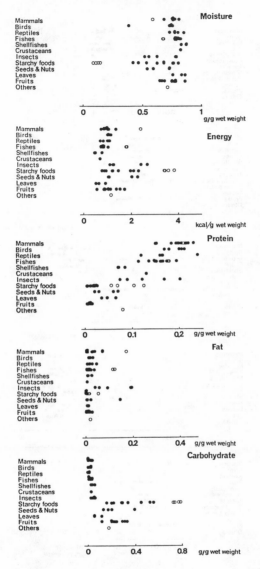

Fig. 1. Moisture, energy, and major nutrient content by food category.
●: local foods, ○: purchased foods.

were stable through the food categories, and variation was also small within each food category. Several samples of shellfish, seeds and nuts, and leaves showed high Mg concentrations, and shellfish and sago flour showed low K concentrations. Among foods of vertebrate animal origin

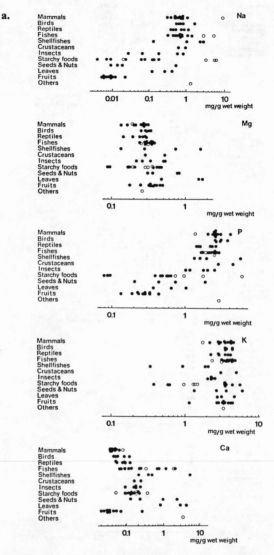

Fig. 2 a.   Na, Mg, P, K, and Ca concentrations by food category.
             ●: local foods, ○: purchased foods.

       b.   Al, Mn, Fe, Cu, Zn, and Sr concentrations by food category.
             ●: local foods, ○: purchased foods.

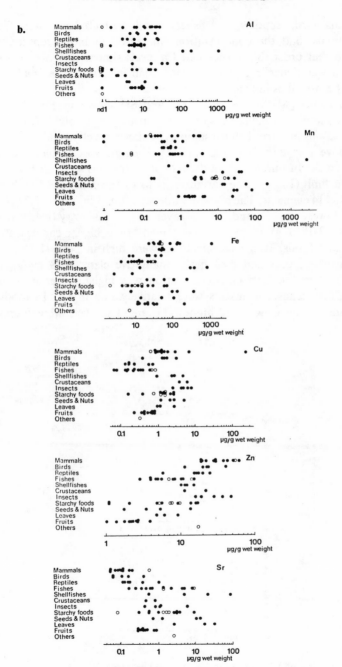

(mammals, birds, reptiles, and fishes), the lowest Ca levels were found in mammals, and the concentrations increased in birds, reptiles, and fishes, in that order. Shellfishes had Ca levels similar to those of fishes. Thus, Ca was richer in aquatic animals than in terrestrial animals. Calcium concentrations in fruits were low.

Six elements (Al, Mn, Fe, Cu, Zn, and Sr) shown in Fig. 2b were usually contained at concentrations up to 1 mg/g wet weight. Aluminum concentrations in three shellfish samples (a mangrove clam and two clams) were above 100 μg/g (117, 202, and 1112 μg/g). Insects had the next highest levels. In purchased foods, Al concentrations were all below the detection limit (1μg/g). The variation in Mn concentrations was largest among the 14 elements. The highest value was found in clams (2.77 mg/g) and the lowest in pork and the fat layer of the cassowary (below detection limit, 0.01 μg/g). Plant foods contained Mn at higher concentrations than animal foods. Iron concentrations were high in pork (1.11 mg/g), shellfish (clam: 0.29 and 0.83 mg/g, mangrove clam: 0.38 mg/g), and cycad (0.36 mg/g). Copper concentration was high in deer liver (0.23 mg/g). Crustaceans and insects were relatively rich in Cu. In foods of vertebrate animal origin, the highest Fe and Cu levels were found in

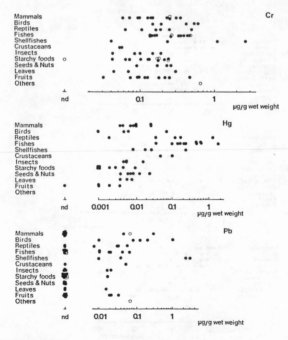

Fig. 2c.   Cr, Hg, and Pb concentrations by food category.
    ●: local foods, ○: purchased foods.

mammals; the concentrations decreased in birds, reptiles, and fishes, in that order. Zinc was present at higher levels in animal foods than in plant foods. Zinc concentrations in vertebrate animal foods showed a trend similar to Cu and Fe concentrations: the highest levels were found in mammals and the lowest, in fish. And Zn concentrations were high in insects and low in shellfish. These imply that Zn was more abundant in terrestrial animals than in aquatic animals. As in the case of Ca, the highest Sr concentrations in vertebrate animal foods were found in fish and the lowest levels, in mammals. Strontium concentrations in shellfish were high; Sr was more abundant in aquatic animals than in terrestrial animals. Strontium concentrations in plant foods were of the same order as in aquatic animals.

Three elements (Cr, Hg, and Pb) shown in Fig. 2c were usually contained at low levels, below 1 $\mu g/g$ wet weight. Two clam samples showed above 1 $\mu g/g$ concentrations in Cr and Pb. Plant foods contained less Hg than animal foods. Among animal foods, fish showed the highest levels of Hg and reptiles the next highest levels.

COMPOSITIONAL CHARACTER

Since purchased foods had different compositional character from local foods, as mentioned above, these were excluded in principal components analysis.

When the values of wet weight basis and moisture content were used as variables, six factors with eigenvalues over 1.0 were extracted, and 84% of variation was explained by these six factors (Table 1). The factor loading matrix after varimax rotation is shown in Table 2. Factor 1 had positive loadings on protein, Na, P, Zn, and Hg and negative loading on carbohydrate; factor 2 had positive loadings on Mg, Ca, Mn, and Sr; factor 3 had positive loadings on energy and fat and negative loading on moisture; factor 4 had positive loadings on Al, Cr, Fe, and Pb; factor 5 had positive loadings on Fe and Cu and negative loading on Hg; and finally, factor 6 had positive loadings on Mg and K.

When dry weight basis values were used for the purpose of controlling the variation in moisture content of each food, five factors with eigenvalues

Table 1. Eigenvalues and Contribution to Variation of Six Factors Extracted by Principal Components Analysis for Values on a Wet Weight Basis

	Factor 1	Factor 2	Factor 3	Factor 4	Factor 5	Factor 6
Eigenvalue	5.260	3.432	2.934	1.684	1.599	1.041
% of variation	27.7	18.1	15.4	8.9	8.4	5.5
Cumulated %	27.7	45.8	61.2	70.1	78.5	84.0

Table 2. Factor Loading Matrix after Varimax Rotation

Item	Factor 1	Factor 2	Factor 3	Factor 4	Factor 5	Factor 6
Moisture	0.304	0.073	-0.837	0.017	-0.149	0.220
Energy	-0.148	-0.072	0.978	0.018	0.079	-0.060
Protein	0.849	-0.369	0.046	0.202	-0.010	0.064
Fat	0.240	0.041	0.880	0.081	-0.001	0.135
Carbohydrate	-0.860	0.051	0.181	-0.165	0.063	-0.293
Na	0.830	0.079	-0.007	0.342	-0.112	-0.173
Mg	-0.047	0.623	-0.071	0.138	0.177	0.637
Al	0.225	-0.013	-0.036	0.721	0.094	-0.327
P	0.925	-0.068	-0.058	0.126	0.064	0.167
K	0.245	-0.140	-0.060	-0.092	-0.102	0.883
Ca	0.120	0.951	-0.031	0.053	-0.058	0.029
Cr	0.092	0.112	0.108	0.823	-0.161	0.207
Mn	-0.366	0.742	0.034	0.111	0.392	-0.169
Fe	0.070	0.089	0.168	0.569	0.608	-0.182
Cu	0.120	0.186	0.054	-0.046	0.877	0.022
Zn	0.754	-0.154	0.037	0.061	0.474	0.003
Sr	-0.223	0.917	-0.056	0.128	0.021	-0.043
Hg	0.600	0.116	-0.158	0.049	-0.615	-0.006
Pb	0.270	0.141	-0.035	0.676	0.072	0.043
% of variation	23.8	15.7	13.4	11.9	10.7	8.5

over 1.0 were extracted. As the eigenvalue of the sixth factor showed a near-1.0 value, 0.984, varimax rotation was conducted including this factor. The principal components structure was identical to wet weight basis values.

When nutrient density calculated by the concentrations in energy units (1000 kcal) for trace elements was used as the variable, four factors with eigenvalues over 1.0 explained 77.8% of variation. When factor 5 (eigenvalue, 0.864) was included, 84% of variation was explained. In the factor loading matrix after varimax rotation with these five factors, factor 1 had large loadings on Na, P, Zn, and Hg; factor 2, on Ca, Mn, and Sr; factor 3, on Al, Cr, Fe, and Pb; factor 4, on Mg and K; and factor 5, on Cu.

As there was no notable difference in the factor loading matrices through these three principal components analyses, factor scores for each food were calculated using the wet weight basis data and plotted in Fig. 3. Factor 1 divided animal foods and plant foods clearly: all of the plant foods except leaves of the tree called *bedum* (*Ormocarpum orientale*) had negative scores, and all of the animal foods except one of two clam samples had positive scores. Factor 2 scores of fish and shellfish were higher than those of mammals, birds, and reptiles in animal foods, and in plant foods those of some leaves and nuts were high.

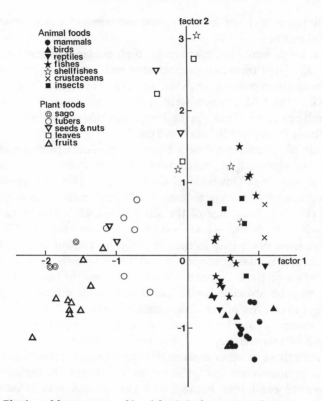

Fig. 3. Plotting of factor scores of local foods in factor 1 and factor 2.

DISCUSSION

In the results of principal components analysis with the values of wet weight basis as variables, moisture represented large negative loading in factor 3 (Table 2). In this factor energy had large positive loading. The highly significantly negative correlation observed between water and energy contents ($r = -0.88$) is reflected in the loading of this factor. Hudson *et al.* (1980) also reported negative correlation between energy and water content of Gambian foods and stressed the importance of water content of foods in predicting the energy content. It is owing to this relationship between water and energy content that a similar factor loading matrix was obtained in dry weight basis concentration and nutrient density controlled by water and energy content, respectively. The fact that no difference was found between the principal components structure of wet weight basis and that of dry weight basis indicated that the compositional character was not basically changed whether water content was controlled or not.

Animal foods were characterized by high contents of protein, Na, P, Zn, and Hg. All of these items represented high positive loading in factor 1 in principal components analyses. On the other hand, in this factor, carbohydrate and Mn, which were relatively abundant in plant foods, had negative loadings. Thus, factor 1 was regarded as the factor dividing animal foods from plant foods (Fig. 3).

In foods of vertebrate animals, there are gradients from mammals to fish in some elemental concentrations; mammals showed highest levels on Zn, Cu, and Fe and lowest levels on Ca and Sr. This trend in elemental concentrations of vertebrate animals is reflected in the level of scores in factor 2 (Fig. 3). The range of the score is $-1.43--0.64$ in mammals, $-1.27--0.55$ in birds, $-0.98--0.15$ in reptiles, and $-0.57-1.49$ in fish. There is no obvious difference in moisture content among the four categories, and similar gradients in vertebrate animals were also found in the profiles of these five elements on a dry weight basis. Thus, these features may be associated with the difference in muscle composition according to the class of vertebrate animals, and the difference in trace element concentrations by the kind of muscle. The fact that the scores in factor 2 of two terrestrial moving monitor lizards are lower (-0.98, -0.75) than those of other aquatic moving reptiles (between -0.56 and -0.15) does not disagree with this notion (of course, the different living environmental conditions between the two groups may influence the composition). Higher Zn concentrations in highly oxidative dark muscle with a large proportion of slow-twitch fibers were reported (Swift and Berman, 1959; Cassens et al., 1963; O'Leary et al., 1979; Maltin et al., 1983). To prove this hypothesis, further comparative studies in the domain of biochemical physiology are necessary.

Shellfish, especially two samples of seawater clams, had high concentrations of Al, V, Cr, Mn, Fe, Ni, Cd, and Pb. It is well known that bivalves, being non-mobile filter feeders, accumulate a variety of elements from their estuarine environment; thus, they are one of the most reliable indicators for heavy metal contamination from anthropogenic sources (Phillips, 1977; Harris et al., 1979; Genest and Hatch, 1981; Leonzio et al., 1981; Eisenberg and Topping, 1984). Although sources of contamination were scarce in the region where our samples were collected, the accumulation was comparable to or higher than reported values from industrialized regions (Harris et al., 1979; Genest and Hatch, 1981; Leonzio et al., 1981; Eisenberg and Topping, 1984; Favretto and Favretto, 1984a; Di Giulio and Scanlon, 1985). Cohen (1985) defined the bioaccumulation factor for marine organisms as the ratio of the concentration of an element in those organisms to its concentration in the waters in which they live. The average values of the bioaccumulation factors he obtained for fishes, crustaceans, and molluscs were based on large number of samples (about 10000 samples of 200 different species) from a wide variety of locations (198 sites around the coastal United States). In his data file, V, Cr, Mn, Ni, Cu, Zn, molybdenum (Mo), and Cd showed higher values in molluscs than in fish and crustaceans. However, the factors of Hg and selenium (Se) were highest in fish. In our results, Hg concentrations in shellfish were not so high as those in fish (Fig. 2c).

Insects contained relatively high levels of fat and Zn (Figs. 1 and 2–b). Oliveira *et al.* (1976) measured contents of major nutrients and trace elements in four species of insects (termite, larvae of two species of moth, and weevil larva) consumed in Angola, and reported high fat and Zn contents as in our case (on a dry weight basis). It is very impressive that the weevil (*Rhyncophorus phoenicis*) larva they measured belongs to the same genus as the sago grub in our samples.

Local starchy foods, sago flour and tubers, are staple foods of the Gidra people (Chapter 9 in this volume). Sago flour had a different character from that of tubers in the results of principal components analysis (Fig. 3). The differences in major nutrient composition between sago flour and tubers were already reported in Chapter 7 of this volume. Even for trace element concentrations, Na, Al, and Fe were higher and Mg, P, and K were lower in sago flour than in tubers.

Of ten samples of fruit, nine were bananas, which are also one of the staple foods of the Gidra people. Chemical compositions of bananas were reported by many researchers (for Papua New Guinea, Hodges *et al.*, 1950; Hispley and Kirk, 1965; Bailey, 1968; Farnworth, 1974; for UK, McCance and Widdowson, 1960; for the USA, Gormican, 1970; Cowgill, 1981; Pennington *et al.*, 1986; for Japan, Teraoka *et al.*, 1981). Compared with the values in these studies, our results are low in Na and high in Al and Fe contents.

Some leaf and nut samples had larger factor scores on factor 2 than other staple plant foods (Fig. 3). Leaf samples were rich in Mg. In particular, abika and *bedum* had high Mg concentrations (1.75 and 1.49 mg/g, respectively) (Fig. 2a). In plant foods, only these two leaves contained V over the detection limit. It is known that Mg is transported and stored in leaves and functions in plants as a constituent of chlorophyll (Wilkinson *et al.*, 1987). Teraoka *et al.* (1981) reported that V content was relatively high in leaf vegetables. Nickel was also detected in these two leaves (abika and *bedum*) and two kinds of nuts, i.e. cycad and lotus. Cataldo *et al.* (1978) found that, following its absorption by the root, Ni was highly mobile in plants, with leaves being the major sink in the shoots during vegetative growth. In addition, Ni, unlike most other trace elements, was readily accumulated in seeds (Mishra and Kar, 1974). However, Ni was not detected in the present samples of jointfir, either in nuts or in leaves.

In the present chapter, attention is paid to compositional character of local foods, which are much more numerous than purchased foods in the senses, of both amounts consumed by the Gidra and numbers of samples analyzed in this study. Among local foods, animal foods are distinguished from plant foods particularly in factor 1, which showed large positive loadings on protein, Na, P, Zn, and Hg contents and negative on carbohydrate content. As for major animal protein sources, four groups are placed in decreasing order of concentrations of Fe, Cu, and Zn, and in increasing order of those of Ca and Sr, as follows: (1) mammals, (2) birds, (3) reptiles, and (4) fish. This group difference comes largely from the scores of factor 2, which showed positive loadings on Mg, Ca, Mn, and Sr. Three groups of plant staples, i.e. sago flour, tubers, and bananas, are distinguished from each other in the compositional character, at least from scores of factors 1 and 2. Finally, five food groups which are not staples, i.e. insects, shellfish, crustaceans, seeds and nuts, and leaves, had their own characteristics.

Food consumption patterns of the Gidra differed from village to village according to the ecological setting. The fauna and flora available in each village are different and the degree of modernization in each village also differs. These differences are reflected in ability to purchase foods (Chapter 9 in this volume). Proportions of plant staples such as sago flour and garden crops (tubers and bananas), purchased starchy foods (rice and wheat flour), and animal protein sources such as terrestrial animals and aquatic animals markedly vary among the villages. Thus, element intake of the Gidra people varies from one village to another and this inter-village variation of elemental intake will be discussed in Chapter 10.

(*Tetsuro Hongo*)

Chapter 9

Food Consumption and Major Nutrient Intake

A number of researchers have disclosed variations in food consumption and nutrient intake among human groups in Papua New Guinea whose race and ethnicity are the same or very close, but whose living conditions differ ecologically and culturally (Langley, 1950; Malcolm, 1970c: 21–22; Oomen, 1971; Norgen et al., 1974; Dwyer, 1983). Harvey and Heywood (1983) compared food consumption of a local group in Simbu Province for three different periods between 1956 and 1981, and demonstrated drastic changes in conjunction with involvement in the cash economy. These variations and changes are of special importance in reaching an ecological understanding of human nutrition.

This chapter focuses on both change and inter-village diversity in food consumption and nutrient intake among the Gidra villages, as determined by food consumption records in Wonie village in 1971–72 and in four villages—Wonie, Rual, Ume, and Dorogori—in 1981. The present analysis will assess the effects of ecological setting on food and nutrition. In addition, a region-specific interest of this study derives from the fact that the Gidra, especially the inlanders, have heavily subsisted on starch of *Metroxylon* sago, which contains negligible amounts of nutrients other than carbohydrates (Chapter 7 in this volume).

THE SUBJECTS

As mentioned previously in this volume, the Gidra villages can be divided into coastal, riverine, inland, and northern groups; the four groups also vary in degree of modernization, which is typically indicated by frequency of visits to Daru, where local products (garden crops, coconuts, fish, game meat, etc.) are sold in the local market and where imported foods (wheat flour, rice, canned fish, etc.) can be purchased in supermarkets. Because the three-km trip requires only one to two hours' travel by sailing canoe, the coastal (Dorogori) villagers frequently visit Daru; each adult visits an average of once or twice a month. The riverine villagers have possessed engine-powered canoes since the 1960s, and in the 1981–82 survey period two or three canoes from each village went to Daru every two weeks or so, each carrying about ten villagers. In contrast, the people of the inland villages, especially the northern ones, walk long distances to the riverine villages to take passage in the canoes; consequently, their visits to Daru are still infrequent, and transporting goods to market is also fairly difficult.

In Gidra society, the basic meal unit is the household. The people usually eat staple foods, three times per day; in addition, snack-like food consumption occurs fairly frequently. The actual times of the three meals vary depending on conditions. Relatively fixed meal hours are in the morning before 9:00 and in the evening at, or after, sunset.

METHODOLOGY

Food Consumption Survey

The 1971–72 survey in an inland village, Wonie, was done by myself over an eight-month period which included the dry (June to November) and wet (December to May) seasons. The 1981–82 survey was designed to include all the Gidra villages, although the food consumption study was carried out by four of our research team members in four selected villages: Rual (northern) by T.K., Wonie (inland) by myself, Ume (riverine) by T.A., and Dorogori (coastal) by T.I. Each of us stayed in one village for four to six months, mostly in the dry season. Food consumption was measured three times in Wonie: in the dry season of 1971, the wet season of 1972, and the dry season of 1981. In each of the other three villages it was surveyed once, in the dry season of 1981.

Each survey recorded amounts of foods consumed by the members of six to eight households for 12 successive days in the 1971–72 survey and 14 successive days in the 1981 survey. Six to eight households were selected because of the necessity for continuous observation and measurement of the foods transported and consumed. The survey duration of 12 or 14 days was determined by our preliminary observation that menus tended to fluctuate from day to day. This occurred partly because animal foods such as game and fish were caught irregularly and partly because sago-working was usually done once every ten days or two weeks by each household in inland and northern villages, and thus the consumption of sago was largely affected by the timing of the sago-working cycle.

The records were obtained using two different methods of estimating the amounts of foods consumed. In one, all foods stored in each household were weighed twice, first in the early morning of the first day of the survey or on the preceding evening, and again on the evening of the last day or early the following morning. In addition, all food flows from, and to, the household during the period were measured. The other method used was the weighing of all foods just prior to cooking or eating. Sometimes we failed to use the latter method because it was difficult to be present each time cooking or snack-eating took place. In these circumstances cooked foods were weighed or, at a minimum, the members of the household were questioned about the foods eaten. Another problem arose from the fact that when people were outside the village, they ate foods which were obtained or harvested elsewhere. In such cases (which were infrequent), the researcher and the subject individual(s) examined several different-sized samples of the same food and estimated the quantity.

One or more village-mates occasionally joined in the meal of a selected household: and, conversely, one or more members of a household sometimes did not take a meal in their household. When this occurred, the amounts of foods consumed were adjusted for by applying man-value coefficients, as described below. On occasions when all members of a subject household stayed out of the village overnight or for a period of time that included meal hours, the food consumption records of the day(s) in question were excluded from the analysis. This reduced the number of net survey days by approximately 20%. In Ume, a villager's death which occurred during the survey period kept all the villagers away from the village for several days, so that the number of net survey days was less than half of the total survey days.

Food Composition

Seventy-eight different foods (70 locally exploited, in the raw state, and eight purchased) consumed by the Gidra were sampled in our 1980 and 1981–82 survey periods, and their major nutrient compositions (protein, fat, carbohydrates, fiber, ash, food energy, and moisture) and element concentrations were determined (Chapters 7 and 8 in this volume); here, the major nutrients, especially energy, protein, and fat, will be exclusively treated. The samples involved many staple foods whose composition had not yet been described in the literature.

The following results of our study of food composition (Chapter 7 in this volume) were directly relevant to the present analysis. First, the water content of sago flour fell from 44.5% on the day of production to 30% on the 20th day; it was eaten during this period of time, usually within ten days after its production. Thus, the present analysis took the changing water content into account. Second, nine types of banana markedly varied in composition; in particular, there is great variation between plantains and ripe bananas. When the variety was one whose composition had been determined, the known values of nutrients were applied. When the variety was not certain, or when it was known but not sampled

Table 1. Breakdown of 92 Foods by the Type of Food Composition

	N[a]	Amount eaten		Energy taken	
		kg	%	kcal	%
Determined in this study	59	1,582	79.6	2,455,437	87.9
From the literature:					
purchased foods	8	49	2.5	130,266	4.7
From the literature: local					
foods with reliable data[b]	16	344	17.3	200,412	7.2
From the literature: local					
foods without reliable data[b]	9	12	0.6	5,985	0.2
Total	92	1,987		2,792,100	

[a] See the text for the classification of foods.
[b] See the text for the recognition of reliability.

for composition analysis, the average values of nutrients of plantains or of ripe bananas were applied. Similarly, average nutrient values were prepared for several food groups such as taros, land mammals, birds, freshwater fish, and saltwater fish, and applied to the animals and plants which were not sampled.

The composition of the remaining foods could not be determined or estimated by our analysis. So we used values derived from the following eight food composition tables: Hodges *et al.* (1950); Bailey (1968); Murai *et al.* (1958); FAO/USDHEW (1972); Creata *et al.* (1976); McCance and Widdowson (1960); Resources Council (Japan) (1982), and Gormican (1970).

For the calculation of nutrient intake for all foods consumed during the survey periods, food composition tables for 92 foods were prepared. Eleven of these foods were categorized from the average composition of several samples. Table 1 shows the breakdown of the 92 foods (and food categories) according to the type of food composition analysis applied. The local foods whose composition comes from the literature are divided into two types: those with "reliable" data are those foods whose composition was determined from data derived from more than one food composition table and whose values varied to only a small extent. Table 1 demonstrates that the foods whose compositions were determined in our study amounted to 80% in weight eaten and nearly 90% in energy taken. In comparison, the consumption of local foods without reliable data was negligible.

Calculation of Per-Person Nutrient Intake

For comparing the nutrient intake among consumption units, either households or villages, the amounts taken were adjusted to those of a

Table 2. Subjects of the Food Consumption Survey

		Male		Female		
		Married	Unmarried	Married	Unmarried	Total
Rual: northern	(1981)	6	7	8	5	26
Wonie: inland	(1971)	4	3	8	11	26
	(1981)	5	3[a]	6	8	22
Ume: riverine	(1981)	5	6	11	5	27
Dorogori: coastal	(1981)	8	9	8	6	31

[a] Including one breastfed baby less than six months old.

single adult male per day, similar to the common technique (for example, Reh, 1962). Based on the energy requirements and safe levels of protein intake per kg of body weight by sex and age recommended by FAO/WHO (1973), the corresponding average values for the 332 adult males of the Gidra (55.7 kg of body weight on average) were calculated. In addition, the energy requirement of each subject and an estimated safe level of protein intake were calculated from his/her body weight and the per-kg values for his/her age. For a pregnant or lactating woman (with a baby less than six months old), an additional allowance was added. From the above values, each individual's adult male equivalent value was determined. Breastfed babies under six months were omitted from the analysis because their requirements were incorporated in the allowance for lactation of the mother (Reh, 1962).

This adjustment process made the inter-village comparison valid despite the fact that the numbers and age/sex compositions of the subjects in the study villages differed (Table 2).

RESULTS AND DISCUSSION

Energy and Nutrient Intake

The amounts of intake of energy and of four kinds of nutrients, adjusted to per-day values for a single male adult, are shown in Table 3. The wet-season records for the Wonie villagers in the 1971–72 survey are not included in this table for the purpose of comparison. In fact, the nutrient intake was almost identical in the dry and wet seasons, in spite of the differences in the proportions of foods eaten (Ohtsuka, 1983). Although the energy and nutrient intake varies from day to day in each household and from household to household in each village, this paper focuses on the comparison among the villages and between the different periods.

A nutritional assessment is valid for energy and protein, for which

Table 3. Energy and Nutrient Intake per Day per Male Adult

		Energy (kcal)	Protein (g)	Fat (g)	Carbo-hydrates (g)	Fiber (g)
Rual	(1981)	3553	54.3	19.6	776	13.0
Wonie	(1971)	3323	48.3	41.1	663	15.5
	(1981)	3550	68.0	10.0	785	16.8
Ume	(1981)	2980	67.6	54.1	555	11.8
Dorogori	(1981)	3221	73.3	25.0	664	10.0

requirements have been established. Using FAO/WHO nutritional re-commendations (1973), the energy requirement of the Gidra adult male (55.7 kg body weight on average) is 2560 kcal for a moderately active individual; the additional allowance for a very active individual is about 500 kcal. Thus, the Ume villagers' energy intake is below the requirement of the very active, while those of the others exceed it.

The inter-village difference demonstrates that the energy intake of northern (Rual) and inland (Wonie) villagers is high, that of riverine (Ume) villagers is low, and that of coastal (Dorogori) villagers is inter-mediate. The body mass index (BMI, weight (kg)/height (m) squared) of adult males was 19.8 for Rual, 20.7 for Wonie, 21.4 for Ume, and 21.8 for Dorogori villagers; this implies that Rual, in particular, and Wonie people are thinner, and Ume and Dorogori people fatter (Chapter 11 in this volume). Thus, the inter-village difference in body physique is ap-parently irrelevant to differences in energy intake. A conceivable reason for this is difference in energy expenditure. Although we have not analyzed the relationship between the energy intake and expenditure, the villagers' activity patterns may be relevant since the northern and inland people travel on foot over longer distances, especially for sago-working and hunting (see Ohtsuka, 1983). Moreover, they more frequently carry heavy things such as raw sago flour, garden crops, and firewood. The riverine and coastal people, in contrast, tend to work near the village, and the riverine people travel by canoe to transport things.

In comparison with energy intake, protein intake is more variable. Adjusting for the quality of protein in the diet relative to that of milk or eggs while considering the amino acid score of the foods (that is, net protein utilization, NPU), FAO/WHO (1973: 71–73) suggests that about 70% be applied as NPU value to the diets of "poor" countries and 80% to "rich" countries. Because of the high proportion of protein intake from animal origin in the Gidra diet, however, the 80% NPU seems more appropriate. Norgan et al. (1974) set 85% NPU as the average for the protein intake by Kaul and Lufa villagers in eastern Papua New

Guinea. If their setting is appropriate, the 85% NPU also seems applicable to the Gidra diet. When 80–85% NPU is applied, the protein intake values (g) of the Gidra as shown in Table 3 should be converted to 43.4–46.2, 38.6–41.1, 54.4–57.7, 54.1–57.5, and 58.6–62.3, respectively. All of these values exceed the FAO/WHO's safe level of 32 g, which was calculated by taking the body weight of the Gidra into account.

Sources of Energy and Nutrient

The proportion of foods which supply energy and nutrients largely differs from village to village. Table 4 shows the breakdown of sources of energy, protein, and fat when the foods are placed in one of eight categories according to their classification; except purchased plant and animal foods, six of them correspond to their subsistence strategies for the foods obtained locally.

As energy suppliers, sago, garden crops (mainly taro, bananas, and yams), and purchased plant foods are most important. During the 1981 survey period, consumption of sago markedly declined from northern village to coastal village; at the same time, the energy from purchased plant foods showed the opposite trend. It seems likely that purchased foods have been substituted for sago during modernization. However, we must also note the weak resistance to salinity of the sago tree (Nishikawa et al., 1979). The sodium concentration of river water near Ume (riverine) in the dry season was about 100 mg/1, while that of creek water near Rual (northern) and Wonie (inland) was about 10 mg/1 (Ohtsuka et al., 1985; also see Chapter 10 in this volume); this environmental factor has limited the availability of sago in Ume and in Dorogori (coastal), although the degree of the limitation cannot be precisely determined.

The energy intake from garden crops in Wonie increased from 1971 to 1981. The immediate reason was the introduction of varieties of sweet potato from the governmental agricultural office in 1979. As a consequence, the proportion of energy from the sweet potato to that from all garden crops rose from only 0.5% in 1971 to 25% in 1981. According to the informants from several villages, new varieties of garden crops, especially sweet potatoes, were introduced in the last several years in many villages, but their yields were high for only about two years after the introduction. This is supported by the fact that the proportion of sweet potato energy in the villages other than Wonie in 1981 was less than 10%.

Animal protein intake was similar in all five survey records, the proportion of aquatic animals to land animals differing with the locality of each village. In contrast, the intake of protein from plant foods varied. Thus, the total protein intake reflects the level of intake of plant protein, especially that of purchased foods such as wheat flour and rice. This is

Table 4. Adult Male Intake of Energy, Protein, and Fat per Day, Derived from Eight Categories of Foods

	Wild plants	Sago	Coconut	Garden crops	Purchased plants	Land animals	Aquatic animals	Purchased animals
Energy (kcal)								
Rual (1981)	227	1912	101	975	205	82	45	6
Wonie (1971)	117	1827	317	918	0	140	4	0
Wonie (1981)	2	1551	10	1591	216	179	0	0
Ume (1981)	27	588	417	1237	537	76	95	2
Dorogori (1981)	2	209	194	1216	1326	27	118	130
Protein (g)								
Rual (1981)	9.3	3.4	0.9	11.6	4.6	16.8	7.5	0.2
Wonie (1971)	1.1	3.3	3.1	9.8	0	30.1	0.8	0
Wonie (1981)	n.a.	2.8	0.1	22.6	6.2	36.2	0	0
Ume (1981)	1.4	1.0	4.1	16.6	12.9	16.4	15.1	0.1
Dorogori (1981)	n.a.	0.4	2.4	16.5	22.2	4.0	21.3	6.4
Fat (g)								
Rual (1981)	2.2	—	11.0	3.9	0.5	0.8	1.0	0.2
Wonie (1971)	1.0	—	34.7	3.6	0	1.6	0.1	0
Wonie (1981)	n.a.	—	1.0	5.6	0.7	3.5	0	0
Ume (1981)	0.2	—	45.8	4.0	1.3	0.8	1.9	0.1
Dorogori (1981)	n.a.	—	12.0	3.1	2.7	1.0	1.4	4.8

n.a. Negligible amount.

related to the difference in protein content per unit of energy: wheat flour and rice contain respectively 3.0 g and 1.6 g protein per 100 kcal energy, while the corresponding values of tubers, bananas, and sago flour are 1.46 g, 0.88 g, and 0.18 g (Chapter 7 in this volume). Among the wild plants which supplied a significant amount of protein for Rual villagers, lotus seed (*Nelumbo nucifera*) was the most important; this plant luxuriates in freshwater swamps, which are most plentiful in the area of the northern villages (including Rual). The increase in protein intake of Wonie villagers from 1971 to 1981 was largely due to the introduction of the sweet potato, which accounted for one-fourth of the total protein intake from garden crops in 1981, the same as its proportion of the energy intake.

Of the fat-supplying foods, the coconut is extremely important. Fat intake from foods other than coconut varies only a little among the five survey records. But a drastic change in coconut consumption occurred in Wonie during the ten years between the two studies. This was caused by increased predation by giant rats (*Rattus rattus*) which gnawed the young fruits of standing trees. Although neither we nor the villagers knew any reason for the rapid population increase of the rats in the inland part of the Gidraland, the rat is well known as a coconut pest in Papua New Guinea (Purseglove, 1972: 469–470; Bourke, 1982).

Diversity versus Change

Strictly speaking, assessment of change in food consumption among the Gidra is possible only by comparing the two records for Wonie. This comparison is typical in showing the introduction of purchased foods and the enlargement of horticulture mainly due to the introduction of sweet potato varieties. Correspondingly, exploitation of sago has declined.

These differences are almost identical to those observed in 1981 records for the four villages. Both differences, over time and among villages in the same year, are judged to reflect the degree of modernization. In other words, the food consumption pattern of each village has changed and continues changing, with substantial inter-village variation among the Gidra population.

CONCLUSION

The above discussion has highlighted the factors determining food consumption and nutrient intake of the Gidra. The relation can be schematically illustrated as in Fig. 1, which classifies the determining factors into three: environmental availability, subsistence technology, and purchasing ability.

Environmental availability determined the degree of dependence on

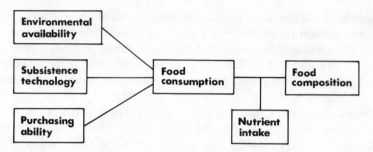

Fig. 1. Schematic diagram showing the relationship between the ecological factors and the food and nutrition.

sago, the relative proportions of land animals and aquatic animals as sources of animal protein, and utilization of region-specific resources such as the lotus seed in Rual village. The decrease in coconut consumption among Wonie villagers is another of the patterns related to environmental availability.

Subsistence technology can be changed by the influence of the outer world through modernization. The introduction of high-yielding varieties of sweet potato is one example; another is introduction of the shotgun and the fish net, both of which were first used about 20 years ago in the area and have become popular especially in the riverine and coastal villages, and have increased efficiency and purchasing ability of the people. The end result is gradual replacement of traditional sago as a plant staple with such purchased foods as wheat flour and rice. As a result, the nutrient intake has also changed, as seen in the increased protein intake per unit of energy (cf. Dennet and Connell, 1988).

In nutritional-ecological studies, it is indispensable to clarify food consumption and nutrient intake in the light of various ecological conditions. This paper has presented the diversity and change of food consumption in a single population with only 1850 people and has analyzed the causal relationships in an ecological framework. It can be concluded that the food consumption and the resultant nutrient intake of the local population tend to diversify due to environmental conditions and also tend to change over time during the process of modernization.

(*Ryutaro Ohtsuka*)

Chapter 10
Intake of Micronutrients

Energy and/or protein deficiency resulting from restricted availability of foods has been evoked as the main nutritional problem in relation to the low carrying capacity of traditional societies. There is, however, a possibility that deficient or excess intake of trace elements and imbalances between the elements occur even where energy and/or protein intake is adequate. The significance of nutritional status of trace elements has become a matter of concern in relation to human health in the developed countries. However, the exact features of trace element intake have been reported only infrequently for traditional populations (Cresta *et al.*, 1976; Ross *et al.*, 1986). In these societies, artificial contaminants which are produced by modern technological activities are thought to be limited compared with the developed countries.

Average levels of energy and protein intake of the Gidra measured by our dietary surveys in 1981 exceeded the FAO/WHO nutritional recommendations for energy requirements and safe protein levels (Chapter 9 in this volume). Food consumption surveys in four selected villages which differ ecologically from one another showed quite conspicuous differences in the composition of foods consumed: the contribution to the daily energy intake of plant staples such as sago flour, garden crops (tubers and banana), and purchased starchy foods (rice, wheat flour, and biscuits) and the contribution to the daily protein intake of terrestrial animals (mammals and birds) and aquatic animals (fish and shellfish) varied markedly among the villages. The difference was related to the environmental availability, subsistence technology, and purchasing ability of each village (Chapter 9 in this volume).

Elemental compositions of the food consumed by the Gidra were reported in Chapter 8 in this volume, in which foods consumed were categorized into 12 groups; the elemental compositions differed from group to group. Thus, elemental intake of the Gidra is expected to vary from one village to another. In the present chapter, intake quantities of 17 elements (Na, Mg, Al, P, K, Ca, V, Cr, Mn, Fe, Ni, Cu, Zn, Sr, Cd, Hg, and Pb) from foods and drinking water are estimated on a village

101

basis; Hg intake is discussed in relation to its major source from fish consumed and also to its concentration in human scalp hair (Chapter 16 in this volume).

METHODS

Food consumption records of this chapter are same as those treated in Chapter 9. The total amount of each foodstuff consumed by all the members of the subject households in the survey period was adjusted to the value for a single adult male per day, using the energy-based man-value coefficients. The daily intake quantities of elements from foods were calculated by applying element concentrations in each food item measured in our laboratory (see Appendix). Element concentrations of some foods, i.e. cassava, pumpkin, pineapple, papaya, and coconut (flesh and liquid in dry and immature fruits), which were sampled in our 1986 survey period and analyzed, were not included in the data of Chapter 8 but in the analysis of this chapter. Therefore, foods recorded during the food consumption survey, those whose element concentration was measured in our laboratory was 97.5% (on the basis of weight consumed) in Rual, 99.6% in Wonie, 94.3% in Ume, and 93.7% in Dorogori. For foods not sampled, concentrations reported by other authors (McCance and Widdowson, 1960; Gormican, 1970; Hamilton and Minski, 1972/1973) or average values for each food category were applied. Food samples whose concentrations were below the detection limit were excluded from the calculation of element intake.

Drinking water samples were collected from the four villages in polyester bottles precleaned with nitric acid solution (for details, Ohtsuka et al., 1985). In Rual and Wonie, people fetch water from creeks or swamps. Ume villagers utilize water of the nearby river, and Dorogori villages, draw water from artificial wells located 50–150 m from the beach. Elemental concentrations were measured by Inductively Coupled Plasma Atomic Emission Spectrometry. Daily water intake was estimated as 1000 ml based on our observation, and the individual daily intake of elements from drinking water was calculated.

For comparison of elemental intake levels in the Gidra with the levels in developed countries, the average values in 50 reports from Japan, the USA, and European countries were consulted for the range of ordinary intake levels.

ELEMENT INTAKE FROM DRINKING WATER

Element concentrations in drinking water for the four villages are shown

Table 1. Element Concentrations in Drinking Water (mg/l)

	Rual	Wonie	Ume	Dorogori	Detection limit
Na	2.47	5.63	48.7	194	0.02
Mg	0.383	0.646	30.6	148	0.005
Al	0.017	0.028	0.012	0.033	0.01
P	0.036	0.025	0.064	0.466	0.01
K	0.108	0.112	2.75	16.4	0.1
Ca	0.160	0.198	51.7	86.5	0.005
V	0.006	n.d.	0.017	0.087	0.005
Cr	0.008	0.003	0.010	0.030	0.003
Mn	0.006	0.005	0.005	0.008	0.005
Fe	0.049	0.012	0.006	0.016	0.005
Ni	n.d.	n.d.	n.d.	0.044	0.02
Cu	0.001	0.002	0.005	0.012	0.001
Zn	0.001	0.003	0.001	0.003	0.001
Sr	0.002	0.003	0.267	1.71	0.005
Cd	n.d.	n.d.	n.d.	0.008	0.005
Pb	n.d.	n.d.	n.d.	0.082	0.05

in Table 1. The figures in the table are arithmetic means of three samples collected in September, November, and December, 1981. The well water of Dorogori village contained extremely high levels of Sr, Ca, Mg, K, and Na compared to the creek or swamp water of the inland villages, Rual and Wonie: the levels of these elements in Dorogori water were 700, 500, 300, 150, and 50 times as high, respectively, as the inland levels. The river water utilized in Ume village contained these elements at intermediate levels between Dorogori village and the two inland villages. Phosphorus, V, Cr, Ni, Cu, Pb, and Cd concentrations in the well water of Dorogori were also higher than those in the water collected in the other villages.

The high Na, Mg, K, Ca, and Sr concentrations in the Dorogori well water can be explained by the mixing of sea water, because the wells are located only 50–150 m away from the beach. The well water's concentrations of these elements are between 2% and 20% of those in sea water (Yamagata, 1977). Concentrations of these five elements in the river water utilized in Ume village show levels intermediate between Dorogori well water and inland creek or swamp water. As the river flows upward when the tide rises at Ume village, water fetched from the river is mixed with sea water. Thus, in Dorogori, contribution of the drinking water to total Sr intake reaches 35.6%, and that of its contribution to Na, Mg, and Ca intake are the highest among the four villages. In Ume, the contribution of the drinking water to intake of these four elements is the second highest. Potassium intake from foods is high compared with the other four elements; the contribution of drinking water to total K intake is negligible (0.3%) even in Dorogori.

In general, sea water contains V, Cr, Ni, Cd, and Pb at levels of about 2, 0.2, 7, 0.1, and 0.03 $\mu g/l$, respectively (Yamagata, 1977). Against this standard, the concentrations of these elements in the well water of Dorogori village were extremely high. In particular, Pb was more than 2500 times higher in the well water. There is a possibility that the sea water near Dorogori village is contaminated by these elements; however, the source of these contaminants is not clear. Our study (Chapter 8 in this volume) indicated that seawater clams collected in Dorogori accumulated these elements at levels comparable to or higher than those in clams caught in industrialized regions of the world (Harris et al., 1979; Genest and Hatch, 1981; Leonzio et al., 1981; Eisenberg and Topping, 1984; Favretto and Favretto, 1984a; Di Giulio and Scanlon, 1985).

COMPARISON OF ELEMENT INTAKE AMONG VILLAGES

Table 2 shows daily intake of trace elements from foods of the four Gidra villages; the range of the intakes reported in Japan, the USA, and European countries are also shown for comparison (discussed later). In this table, Hg intake reported in Chapter 17 is used.

Total daily intake levels from foods and drinking water in the four villages are shown in Figs. 1-3. The five elements (Na, Mg, P, K, and Ca) in Fig. 1 are consumed in g (gram)-order quantities; the six elements (Al, Mn, Fe, Cu, Zn, and Sr) in Fig. 2 at the mg level, and the six elements (V, Cr, Ni, Cd, Hg, and Pb) in Fig. 3 at the μg level. As the consumption of table salt is a large fraction of total Na intake (55% to 82%), Na intake from salt is shown separately from that from other foods. In comparison among the four villages, the difference between the highest and the lowest exceeds two times for Na (Fig. 1), Al and Fe (Fig. 2), and V, Ni, Cd, Hg, and Pb (Fig. 3).

The highest Na intake, that in Dorogori (1.8 g/day), is attributed to relatively high consumption of salt and salty purchased foods and to extremely high Na concentration in the drinking water. Purchased foods provide 40.7% of Na intake from foods (except for salt) and 13.5% of the total Na intake. Moreover, the abundance of seawater fish and coconut attributed to the location of this village, on the coast, increases Na intake. Each of these two kinds of foods provides about 20% of Na intake from foods. Daily urinary Na excretion of adult males in Dorogori was the highest among the four villages, reflecting the high Na intake (Chapter 15 in this volume).

In Wonie, Na intake was the lowest among the four villages, reflecting the lowest salt consumption.

Intake levels of Al and Fe were highest in Rual and lowest in Dorogori.

Table 2. Trace Element Intake from Foods in Four Villages, Compared with the Range Reported in Japan, USA, and European Countries

	Rual	Wonie	Ume	Dorogori	Japan[a]	USA and European countries[b]
Na (g)	1.14	0.72	1.42	1.59	3.2 – 7.2	1.9 – 7.2
Mg (g)	0.59	0.69	0.64	0.63	0.19– 0.33	0.23– 0.44
Al (mg)	25.8	24.1	16.9	10.9	4.5 – 5.7	2.3 – 27
P (g)	1.30	1.70	1.69	1.96	0.93– 1.4	1.0 – 2.0
K (g)	5.32	7.82	5.24	6.31	2.0 – 3.2	2.2 – 4.9
Ca (g)	0.44	0.46	0.45	0.40	0.47– 0.73	0.62– 1.5
V (μg)	4.7	1.9	40.9	25.9	230	10 – 28
Cr (μg)	263	332	228	259		28 –320
Mn (mg)	30.6	24.2	17.2	25.0	2.8 – 8.7	0.9 – 6.1
Fe (mg)	97.1	60.1	55.7	32.1	6.1 – 17	5.2 – 23
Ni (μg)	225	1.4	85.9	12.3	190 –280	130 –180
Cu (mg)	2.98	3.65	3.00	2.68	0.91– 3.6	0.40– 3.4
Zn (mg)	12.1	16.2	16.4	13.5	6.5 – 12	6.7 – 18
Sr (mg)	3.53	4.28	3.76	3.09	1.0 – 2.3	0.86– 1.9
Cd (μg)	0	0	6.0	2.1	35 – 59	13 – 64
Hg (μg)	16.4	8.0	21.6	81.3	4.9 – 20	2.9 – 20
Pb (μg)	1.79	3.34	8.71	43.1	57 –220	60 –320

Sources: [a]Goto et al., 1972; Horiguchi et al., 1978; Inoue et al., 1985; Ishida et al., 1988; Ishigure et al., 1985; Ishimatsu, 1988; Iwao, 1977; Iwao et al., 1981; Kanazawa and Muto, 1984; Kotake et al., 1981; Kunisaki et al., 1984; Murai et al., 1985; Nagahashi et al., 1985; Nishihara et al., 1979; Shirai, 1988; Shiraishi et al., 1986; Shishido and Suzuki, 1974; Suzuki and Lu, 1976; Tanaka et al., 1983; Teramoto et al., 1987; Teraoka et al., 1981; Yomota et al., 1987; Yoshida and Ikebe, 1988; [b]Anderson and Kozlovsky, 1985; Buchet et al., 1983; Byrne and Kosta, 1979; Calkins et al., 1984; Clemente et al., 1977; Evans et al., 1985; Farre and Lagarda, 1986; Flyvholm et al., 1984; Freeland-Graves et al., 1980; Gibson et al., 1983; Gormican, 1970; Greger, 1985; Hamilton and Minski, 1972/73; Hazell, 1985; Holden et al., 1979; Kelsay et al., 1988; Mahaffey et al., 1975; Marsh et al., 1988; Myron et al., 1978; Pennington, 1987; Schelenz, 1977; Sherlock et al., 1982; Spring et al., 1979; Taber and Cook, 1980; Tylavsky and Anderson, 1988; Varo and Koivistoinen, 1980; Wenlock et al., 1979.

In particular, Fe intake in Rual was extremely high. The amount of consumption of sago flour, which contains 2.0 mg of Al and 4.5 mg of Fe per 100 g, was reflected in the levels of intake of these two elements among villages; the contribution of sago to total energy intake was the highest in Rual (54%) and the lowest in Dorogori (6.5%). The extremely high Fe intake in Rual, however, resulted from not only sago but also other foods rich in Fe, cycad (*Cycas circinalis*) and lotus (*Nelumbo nucifera*) seeds. Iron intake from these two wild seeds was 31.3 mg (32.2%). In Ume, Fe intake from shellfish was as high as 13.5 mg (24%), which raised total

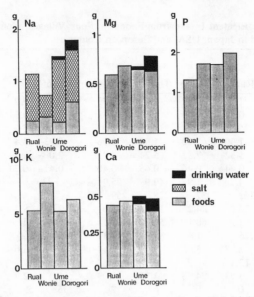

Fig. 1. Daily intake of Na, Mg, P, K, and Ca from foods and drinking water in the four villages.

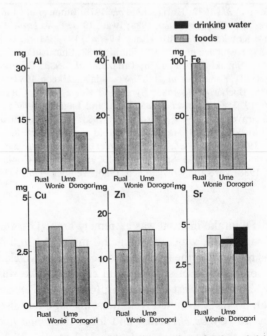

Fig. 2. Daily intake of Al, Mn, Fe, Cu, Zn, and Sr from foods and drinking water in the four villages.

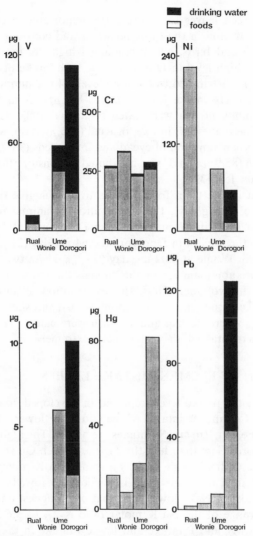

Fig. 3. Daily intake of V, Cr, Ni, Cd, Hg and Pb from foods and drinking water in the four villages.

Fe intake in this village to the same level as that in Wonie, in spite of its low sago consumption compared with Wonie.

Vanadium and Cd intake was highest in Dorogori village because of the high concentrations in the drinking water, its contribution to total intake was 77% and 79%, respectively. On the other hand, intake of these two elements from foods was highest in Ume since shellfish for both elements and green leaves for V were the major source of the intake. High Pb concentration in the drinking water and consumption of foods such as shellfish and birds, which contained high Pb, increased the Pb intake to a very high level in Dorogori. The high intake of Hg in Dorogori (81 μg/day) is attributed to the high levels of consumption of fish in this village, while Hg intake was as low as 8.0 μg/day in Wonie where very small amounts of fish were eaten (Ohtsuka, 1983; Chapter 17 in this volume). The highest Ni intake, in Rual (225 μg/day), was attributed to high levels of consumption of cycad and lotus seeds (98%); the second highest, in Ume (85.9 μg/day), was attributed to consumption of shellfish (61%) and *bedum* leaf (32%).

Although Ca intake from foods differs to a negligible degree among villages (0.40 to 0.46 g/day), the contribution of animal foods differed markedly, from 1.8% in Wonie to 23% in Dorogori. Intake of Sr from animal foods is the highest in Dorogori at 0.64 mg/day (21% of the total) and the lowest in Wonie at 0.02 mg/day (0.5%). These two elements, Ca and Sr, are more abundant in aquatic animals than in terrestrial animals (Chapter 8 in this volume). Thus, the contribution of animal foods to Ca and Sr intake is thought to correspond to fish and shellfish consumption. In Wonie, where no fish and shellfish were eaten, the contribution of animal foods to intake of these elements is extremely low.

EVALUATION OF ELEMENT INTAKE LEVELS

Compared to the intake levels reported in developed countries (Table 2), Na, Ca, Ni, Cd, and Pb intakes of the Gidra are lower, while Mg, Al, P, K, Cr, Mn, Fe, Cu, Zn, and Sr intakes are higher. The highest Na intake level in Dorogori is less than half the Japanese level. On the other hand, K intake in the Gidra is high, and as a result, Na/K molar ratio in the Gidra is lower than 0.5. This ratio is particularly low, 0.16, in Wonie, because the intake of Na is lowest and that of K highest among the four villages. (The range of the Na/K molar ratio is 2.4 to 5.6 in Japanese.) Relatively high Mg and P intake and low Ca intake in the Gidra result in high Mg/Ca (2.2 to 2.6) and P/Ca (3.8 to 6.3) molar ratios compared with those for Japanese, 0.6 to 0.9 and 2.1 to 2.5 respectively. One of the causes of low Na intake in the Gidra is low consumption of salt. Low Na and Ca intake can be attributed to the low consumption of animal foods, especially of dairy products in the case of Ca. The similar low Na and Ca intake has been reported in several groups of vegans (Calkins et al., 1984; Freeland-Graves et al., 1980).

Iron intake of the Gidra is extremely high. Since Mn and Al intake is also high, one of the causes of high Fe intake comes from contamination of food with soil, as was suggested by Moser et al. (1988) in the case of Nepalese high Fe intake (32.7 mg). However, it is also necessary to consider the contribution of some region-specific foods such as sago, wild seeds, and shellfish. Ross et al. (1986) reported that Fe intake in Wosera, Papua New Guinea, was 43.3 mg and that green leaves provided 82.4% of total intake. In the highland population of Ethiopia, daily Fe intake of 200–500 mg is common because of the consumption of teff (Eragrostis abyssinica), a cereal with an exceptionally high Fe content (Fransson et al., 1984; Hofvander, 1968). Because of high Fe intake, gestational anemia rarely occurs in this population (Gebre-Medhin and Birgegard, 1981; Gebre-Medhin et al., 1976). In the Gidra, high prevalence of hookworm infestation in lowland Papua (Ewers and Jeffrey, 1971) must be taken into account in studying the significance of high Fe intake. High intake of Al and Mn in the Gidra is also attributed to sago and shellfish consumption; these provide 55–65% and 45–70%, respectively, of total Al and Mn intake.

Nickel, Cd, and Pb intake levels in the Gidra may be underestimated because they are calculated based on only the foods that contain these elements above the detection limit. However, the intake of these elements is certainly lower than that in industrialized countries, except for Ni in Rual and Pb in Dorogori. In Dorogori, Pb intake from foods is 43.1 $\mu g/$ day; it increases to 125 $\mu g/day$ when intake from drinking water is taken into account. This Pb intake level in Dorogori is about 30% of the provisional tolerable weekly intake of 3 mg recommended by WHO (1972). As for Hg, intake in Dorogori village (81 $\mu g/day$) is higher than that reported in fishing communities in the United Kingdom (Sherlock et al., 1982), and this level exceeds the tolerable weekly intake recommended by WHO (1972). This extraordinarily high Hg intake in Dorogori was provided by the meat of a shark which was caught incidentally during the food consumption survey period: Hg concentration in shark meat is very high (1.9 $\mu g/g$), and its consumption supplied about 50 μg Hg/person/day (Chapter 17 in this volume).

High intake of Mg, P, K, Cr, Cu, Zn, and Sr in the Gidra may be related to their high energy intake. Ross et al. (1986) carried out food consumption survey in Wosera, Papua New Guinea, and reported that Cu and Zn intake levels were 2.7 mg and 8.4 mg. Their average energy intake was 1900 kcal, and Cu and Zn intake per 1000 kcal was calculated as 1.4 mg and 4.4 mg, respectively. Although Zn intake in the Gidra was about 15 mg, the level per 1000 kcal (4.4 mg) was comparable with that of the Wosera people. Copper intake in the Gidra was as same as that in the Wosera, and the level per 1000 kcal was 0.9 mg.

The Gidra's diet was characterized by high dependence on plant foods compared with that of developed countries; the contribution of animal foods was only 3.7% to 8.5% of the daily energy intake. As a result, crude fiber intake in the Gidra was high, from 10.0 g to 16.8 g (Chapter 9 in this volume) compared to Japanese, 4.1–5.7 g (Nagahashi et al., 1985). The bioavailabilty of Ca (Allen, 1982; McBean and Speckmann, 1974), Fe (Hallberg et al., 1989; Bindra and Gibson, 1986; Hallberg, 1981) and Zn (Turnlund et al., 1984; Solomons, 1982; Drews et al., 1979) from plant foods is lower than that from animal foods because of the generally lower protein contents of plants and also because they contain inhibitors of absorption such as oxalic and phitic acids and some types of fiber. The level of Ca intake in the Gidra is low compared to the recommended dietary intake of 0.6 g in Japan (Japanese Ministry of Health and Welfare, 1984) and 0.8 g in USA (National Research Council, 1980), and the contribution of animal foods to total Ca intake is low, especially in Wonie (1.8%).

Although Zn intake in the Gidra is comparable to the 15 mg recommended dietary intake in the USA (National Research Council, 1980), the contribution of animal foods to total Zn intake (16.2% to 32.5%) is low compared to other countries. Hazell (1985), in a review on dietary sources and bioavailability of trace elements, reported that the contribution of animal foods in the average UK diet was about 60%. In our study on Japanese young adult women, about half of total Zn intake was provided by animal foods (Ishida et al., 1988). High Fe/Zn molar ratios of more than 2.5, especially in the case of nonheme Fe from plant foods, may also inhibit the intestinal uptake of Zn (Solomons and Jacob, 1981). High Fe/Zn molar ratios in the Gidra (2.8 to 9.4) may indicate lower availability of Zn in this population.

(Tetsuro Hongo)

III. NUTRITION AND HEALTH

Overview

The health status of human beings, either as individuals or in populations, is determined by both genetic and environmental factors. Since the Gidra people have survived in small numbers, forming a unity of marriages and thus of gene pools over generations (cf. Chapter 19 in this volume), it can be judged that their health conditions primarily reflect environmental factors, or, more precisely, the adaptive mechanisms by which they adapt to the immediate environment. Because no statistic data are available for the area, all health or biological indicators were measured by us, some in the field and others in the laboratory with biological specimens (i.e. urine and scalp hair) sampled in the field.

For body physique and composition, adult data (Chapter 11) measured in all of the 13 Gidra villages as well as child and adolescent data measured in four intensively studied villages twice at one-year intervals for assessment of growth (Chapter 13) are compared among villages in relation to subsistence and nutrition. A study of infant growth is connected with rearing behavior on the one hand, and disease and nutrition on the other, although because of the small number of infants in many villages this study was conducted in one village only (Chapter 12).

The measurements of blood pressure, which is less subject to aging effect than in advanced countries, demonstrate inter-village differences in changes with age, and this is discussed in the light of nutritional-physiological adaptation (Chapter 14). Urinalysis focuses on protein nutriture by comparing the content of urinary urea nitrogen, which is a sensitive indicator of protein intake, among villages as well as among sex/age groups (Chapter 15). Element concentrations in scalp hair show marked inter-village differences. Chapter 16 aims to relate the hair micronutrient concentrations to levels of intake, and also to consider their validity in the assessment of human health and nutrition. On the other hand, Chapter 17 analyzes concentrations of mercury (Hg), one of the most hazarous contaminants, in scalp hair and in fish eaten in the area, since markedly high Hg concentrations have been noted in some Papua New Guinea populations.

Chapter 11

Adult Body Physique and Composition

The investigation of various populations of Papua New Guinea has brought to light a spectacular array of cultural, linguistic, and physical diversity. Human biological studies have revealed variation in body physique as a result of the interaction of genetic and environmental differentiation, but largely confined to inter-population variation. According to the reports from 38 groups, the range of mean stature is nearly 20 cm in each sex (148.8–167.8 cm in men and 137.0–156.8 cm in women) and mean weight ranges from 44.3 kg to 60.5 kg in males and from 38.4 kg to 54.0 kg in females. Numerous reports have presented the general features of highlanders, mainlanders, and islanders: highland males and females are the smallest in stature but their weights are rather heavy; mainland peoples are taller and lighter than highlanders; the heaviest are islanders, who are taller than mainlanders (for details, Kawabe, 1986a).

Origin of variation within the regions and locations has been accounted for in Papua New Guinea (Heywood, 1983). Male morphological differences among populations such as Bougainville Islanders (Friedlaender, 1975), eastern highlanders (Littlewood, 1972), and Enga-speaking peoples (in the highlands) (Freedman and Macintosh, 1965) were attributed to genetic difference or filtration of genes, with the suggestion of environmental effects including altitude, population density, land-use pattern, nutrition, and disease.

No study has focused on environmental effects in exemplifying the differences within a population. A human population, defined as a reproductive unit occupying a particular space, fulfills an important role in human ecological study as a basis of adaptation and survival.

A total ecological approach to a particular population will develop a holistic view of the complex relationship among various biological traits (e.g. size, shape, and composition) and the environmental variables (e.g. climate, physiography, nutrition, and culture). This chapter treats intra-population variation of body physique among the Gidra. Since 95% of the married members were born within their land, the Gidra can be regarded as a biotic population (Chapter 20 in this volume). In this chapter, the 13 Gidra villages are grouped into inland/northern and riverine/coastal groups according to their village locality, subsistence activity, nutrition, and degree of modernization.

SUBJECTS AND METHODS

The inland/northern group consists of eight villages, two northern (Rual and Kapal) and six inland (Iamega, Wipim, Podare, Gamaeve, Wonie, and Kuru) villages. Of the five riverine/coastal villages, Ume is along the Binaturi River, Wuroi, Woigi, and Abam are along the Oriomo River, and Dorogori is located on the coast, facing Daru across a strait 3 km in width.

As classified according to the age-grade system of the Gidra (see frontispiece), married men (*miid, nanyuruga,* and *rugajog*) and women (*nanyukonga* and *kongajog*) were selected as subjects, as well as 34 unmarried persons (17 *kewalbuga* and 17 *ngamugaibuga*) judged to be adults on the basis of genealogical data and pregnancy histories. For examination of the aging process, the subjects are divided into adults (*rugajog* men and *kongajog* women) and elders (*miid* and *nanyuruga* men and *nanyukonga* women). After the physically handicapped, persons suffering or recovering from severe illnesses, and pregnant women were excluded, a total of 332 men and 419 women were submitted to analysis, a number which corresponded to about 90% of the total number of adults (362 men and 470 women).

The field work was undertaken twice, one for three months from June to September 1980 and again for six months from July 1981 to January 1982 (Kawabe, 1986b). During the first period, the stature and weight of almost all the Gidra people in the 13 villages were measured. During the second period, an intensive investigation was conducted in four selected villages, Rual, Wonie, Ume, and Dorogori, which included measurements of stature, weight, three circumferences (chest, upper arm, and calf), and two skinfold thicknesses (triceps and subscapular). Anthropometric measurements were conducted in the open air or under a floor built more than 2 m high; even and hard ground was selected because flat and firm floors are lacking in the Gidra houses. All the measurements were done by to avoid inter-observer error (Jamison and Zegura, 1974; Kouchi and Hanihara, 1981).

For indicating body build and/or fatness, the body mass index (BMI: weight (kg) divided by stature (m) squared), also known as the Quetelet index, was adopted, since it has been judged the best index because of its independence of height and high correlation with body fatness (Keys *et al.*, 1972). Body density was estimated from the skinfold thickness using the 20–29 age-group equation of Durnin and Womersley (1974):

Body density $(kg/m^3) = (c - m \log \sum 2skinfolds) \times 10^3$

where c is 1.1525 and m is 0.0687 for men, and c is 1.1582 and m is 0.0813 for women, after consideration of the results in coastal and highland New Guineans by Norgan et al. (1982). The percentage of the body as fat (%fat) was calculated from density using the equation of Brožek et al. (1963). Fat-free mass (FFM) and fat mass (FM) were derived from the %fat and body weight. Arm muscle circumference, muscle area, and fat area were derived from measures of upper arm circumference and triceps skinfold thickness (for details, see Table 3).

The normal or Gaussian distribution hypothesis was tested by computing the Kolmogorov D statistic and its probability (Stephens, 1974), and the skewness and kurtosis were calculated using UNIVARIATE procedure of the computer program of SAS (SAS Institute, 1982). For measuring the degree of heterogeneity among villages, analysis of variance (ANOVA) was used to test the null hypothesis of equality of group means after Bartlett's test of equality (homogeneity) of variances. Duncan's multiple range test at the 5% significance level (Duncan, 1975) was applied to compare each pair of groups (villages). In order to analyze simultaneously more than two variables, multivariate analysis of variance (MANOVA) was performed and Wilks's λ was calculated for testing the equality of group means. Mahalanobis's generalized distance (D^2) was computed to compare each pair of means, testing each equality of means by Hotelling's T^2. For extracting a simplified figure of the D^2 matrix, a cluster analysis was applied (Dixon and Brown, 1979).

RESULTS

Frequency Distribution of Stature and Weight

The data on stature and weight from the whole Gidra population were first examined for whether they represented a normal distribution. In Figs. 1 and 2 the distributions of stature and weight of men and women are shown as histograms in the class intervals of 2 cm and 3 kg respectively. The distribution of stature for each sex, on which the fitted normal curve was superimposed (Fig. 1), accepted the hypothesis of normality since Kolmogorov Ds of men and women were 0.043 and 0.031 respectively ($p \gg 0.05$), and the data showed no significant skewness or kurtosis.

Men's weight also has an approximately normal form: $D = 0.049$ ($p = 0.047$) (Fig. 2). In contrast, the distribution of women's weight is asymmetrical or positively skewed (skewness = 0.911, $p < 0.01$) and is significantly leptokurtic (kurtosis = 1.725, $p < 0.01$); hence the data are not

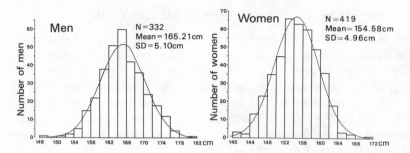

Fig. 1.   Frequency distribution of stature of the Gidra adults and elders.

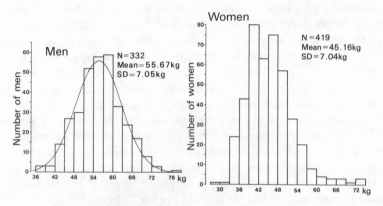

Fig. 2.   Frequency distribution of weight of the Gidra adults and elders.

normally distributed ($D = 0.060$, $p < 0.01$). Normalization was attempted by square root, cubic root, and logarithmic transformation. The logarithmic transformation led to a satisfactory normalized distribution in female weight, while none of the transformed distributions in male weight were normal. Thus, in the following analysis, logarithmic transformation will be used for female weight together with untransformed value when an analysis may be affected by the non-normality of distribution. However, a positive skewness remained significant ($p < 0.01$) even in the distribution of transformed female weight. A skewness test which excluded, one by one, the highest values from the data proved to be not significant when the 12th heaviest subject, 61.5 kg, was removed. Four of the 12 heavy subjects belonged to the inland/northern villages, and the remaining eight lived in riverine/coastal villages; seven of the eight lived in the coastal village, Dorogori.

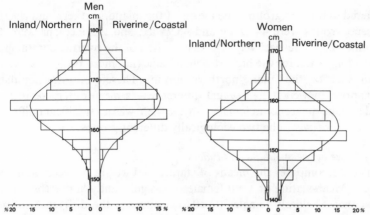

Fig. 3. Frequency distribution of stature between inland/northern and riverine/coastal groups.

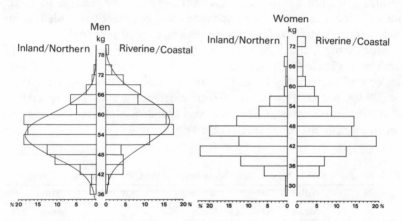

Fig. 4. Frequency distribution of weight between inland/northern and riverine/coastal groups.

The distribution of stature and weight was examined separately for the inland/northern and riverine/coastal groups. The stature distributions proved to be normal for both groups (Fig. 3), each showing a similar shape in both sexes (SD was nearly 5.0 cm in every group). The differences in means, 1.00 cm for men and 0.61 cm for women, were not significant (p > 0.05) as measured by t-statistics. In contrast, the weight distribution varied in the two groups (Fig. 4). Men's distribution, normally shaped, had the same configuration in both groups, while the position of the mean was significantly higher in the riverine/coastal group than in the inland/northern group: 58.05 kg and 54.07 kg respectively. Women's weight differed in the two groups; the means of the inland/northern and riverine/ coastal groups were 44.22 kg and 46.95 kg, and the medians were 44.0 kg and 45.5 kg, respectively. The distribution of the riverine/coastal group had a longer tail for the highest weight value (i.e. more positive skewness) than that of the inland/northern group. Therefore, intra-population variations in weight had sexual differences—a mean difference in males and a shape difference in females—while there was no significant difference in stature between the two ecologically different groups.

Difference between Adults and Elders

Table 1 compares the means of stature and weight between adults and elders. Mean stature of adult females was significantly higher (by 1.56 cm) than that of elderly females ($p < 0.01$), although the difference in mean stature for men, 0.97 cm taller in adults than in elders, was not significant at the 5% level. This difference between adults and elders probably indicates a stature decline with age, although it is impossible to clearly identify since the age of adults and elders was not exactly recorded: adults were approximately 20–50 years old; elders, over 50 years old.

Much more marked, however, is the decrease in weight with age for both sexes (Table 1), showing a male predominance: the weight of elder males was significantly lower than that of adult males (mean difference = 6.29 kg, $p < 0.001$) while the female difference was 4.12 kg ($p < 0.001$). A significant decrease of BMI was also found in each sex, showing, like weight, a male predominance (Table 1).

Table 1. Stature, Weight, and Body Mass Index of the Gidra Adults and Elders

Variable	Adult			Elder			Differ-ence	Total		
	N	Mean	SD	N	Mean	SD		N	Mean	SD
Male										
Stature (cm)	254	165.44	5.00	78	164.47	5.39	0.97^{NS}	332	165.21	5.10
Weight (kg)	254	57.14	6.36	78	50.85	7.07	6.29***	332	55.67	7.05
Body mass index	254	20.85	1.85	78	18.77	2.15	2.08***	332	20.36	2.11
Female										
Stature (cm)	320	154.95	4.97	99	153.39	4.74	1.56**	419	154.58	4.96
Weight (kg)	320	46.13	6.64	99	42.01	7.41	4.12***	419	45.16	7.04
Body mass index	320	19.18	2.32	99	17.84	2.91	1.34***	419	18.86	2.54

^{NS} Not significant; * $p < 0.05$, ** $p < 0.01$, *** $p < 0.001$.

Variation of Stature and Weight among 13 Villages

The mean and standard deviation (SD) of stature and weight were calculated for each of 13 villages, and are listed in Table 2. The SDs are very uniform and small in stature for both sexes (by Bartlett's test); no observed range exceeds the mean plus or minus 3 SDs. The stature data, normally distributed, show small differences between the means, implying that the stature of the villagers is similar and shows a reasonable amount of random variation.

The F-value of stature resulting from ANOVA shows a small but significant difference ($0.01 < p < 0.05$) of group means among 13 villages (Table 2). However, unequal sample numbers in the 13 villages ($\chi^2 = 40.46$, $p < 0.001$ for males; $\chi^2 = 51.64$, $p < 0.001$ for females) may have affected the F-test (Snedecor and Cochran, 1980: 228). Results of Duncan's multiple range test to compare each pair of villages represent nine pairs of significant mean differences in males and eight pairs in females. A noteworthy feature is that the maximum difference was found between geographically neighboring villages for both sexes: 5.62 cm between Gamaeve and Podare in males and 4.46 cm between Abam and Dorogori in females. These differences indicate that due caution should be exercised in representing a population by a mean calculated from a small number of samples, for example, members of a single village.

In comparison with stature, coefficients of variation of weight were larger and the observed range exceeded the mean plus or minus 3 SDs in Iamega for males and in Rual, Kuru, and Ume for females (Table 2). The resultant F-values by ANOVA were 4.70 in men and 7.06 in women; thus the weight means were significantly different ($p \ll 0.001$). Male weight demonstrates a noticeable pattern. The highest mean was that for Ume village; the second highest, for the coastal village, Dorogori. According to Duncan's multiple range test, the five riverine/coastal villages were included in the highest group, although four inland/northern villages, Podare, Wonie, Wipim, and Rual, were also included in it. The lowest group consisted of only inland/northern villages, Rual, Kuru, Kapal, Gamaeve, and Iamega.

Inspection of female weight proves that, in general, mean weight appears to be greater in the riverine/coastal villages and smaller in the inland/northern villages, though the pattern is less clear than in male weight. Weight mean of the Dorogori women was by far the highest, forming the highest group by itself (Table 2) and showing a significant difference from the other 12 female villagers, all of which fall into two wide-range groups of non-significantly different means. Similar results were obtained in the same analysis with logarithmic transformed values.

Table 2. Stature and Weight of the Gidra Adults and Elders in 13 Villages

Village	N	Stature (cm) Mean[a]	SD	Range	Weight (kg) Mean[a]	SD	Range
Male							
Rual	20	164.69^{BCD}	5.15	156.7–175.0	55.18^{ABCD}	6.80	45.0–67.0
Kapal	19	164.59^{BCD}	5.41	155.0–173.7	52.39^{CD}	5.74	42.0–64.5
Iamega	50	164.48^{BCD}	5.08	153.1–176.4	51.43^{D}	5.30	42.0–69.0
Wipim	25	164.39^{BCD}	4.90	147.2–173.7	56.90^{AB}	6.51	38.0–72.5
Podare	16	168.13^{A}	4.51	160.0–175.1	57.19^{AB}	7.59	36.5–68.5
Gamaeve	18	162.51^{D}	6.61	150.8–177.0	52.28^{CD}	5.62	40.5–61.0
Wonie	23	165.54^{ABCD}	4.43	156.9–176.8	57.17^{AB}	5.27	47.5–69.0
Kuru	28	165.05^{ABCD}	4.83	154.2–174.8	53.43^{BCD}	7.05	38.0–65.0
Ume	39	166.73^{ABC}	5.89	157.0–181.4	59.17^{A}	8.92	42.5–78.5
Wuroi	24	164.93^{ABCD}	3.69	156.9–171.3	56.27^{ABC}	6.38	42.5–69.5
Woigi	24	167.63^{AB}	3.42	161.6–175.3	58.19^{A}	5.72	46.0–68.0
Abam	24	163.63^{CD}	5.07	154.5–175.1	57.65^{AB}	5.60	43.0–68.5
Dorogori	22	165.41^{ABCD}	4.95	156.7–174.0	58.32^{A}	7.90	40.0–71.5
Total	332	165.21	5.10	147.2–181.4	55.67	7.05	36.5–78.5
Bartlett's test		15.51 (p = 0.2273)			20.43 (p = 0.0651)		
F-value		2.07 (p = 0.0187)			4.70 (p < 0.0001)		
Female							
Rual	26	153.86^{BC}	4.91	140.4–162.8	46.71^{B}	6.91	33.0–68.5
Kapal	30	153.97^{BC}	4.79	145.3–164.0	42.60^{C}	5.12	34.5–50.5
Iamega	68	155.49^{ABC}	4.34	144.5–163.9	44.43^{BC}	5.53	33.5–56.5
Wipim	28	154.99^{ABC}	4.15	149.2–165.4	46.02^{BC}	6.02	36.5–61.5
Podare	35	153.73^{BC}	4.90	144.2–163.1	43.93^{BC}	6.49	28.0–59.5
Gamaeve	31	153.21^{BC}	5.61	140.9–163.9	42.89^{BC}	6.44	33.5–58.5
Wonie	26	154.62^{ABC}	5.05	146.0–163.5	45.12^{BC}	4.25	37.0–52.5
Kuru	32	153.89^{BC}	5.49	142.2–166.4	42.61^{C}	8.07	30.5–68.5
Ume	41	153.99^{BC}	4.54	146.5–166.8	45.76^{BC}	7.99	33.0–71.5
Wuroi	23	154.73^{ABC}	4.56	145.0–162.4	44.54^{BC}	4.90	36.5–54.0
Woigi	28	156.06^{AB}	4.83	141.8–161.9	44.66^{BC}	5.15	36.0–58.0
Abam	24	152.93^{C}	5.42	142.9–163.7	44.54^{BC}	6.36	33.5–58.0
Dorogori	27	157.38^{A}	5.60	145.9–170.9	55.35^{A}	9.38	34.0–72.5
Total	419	154.58	4.96	140.4–170.9	45.16	7.04	28.0–72.5
Bartlett's test		7.48 (p = 0.8309)			35.29 (p = 0.0005)		
F-value		1.87 (p = 0.0365)			7.06 (p < 0.0001)		

[a] Means with the same letter ($^{A, B}$ etc.) are not significantly different by Duncan's multiple range test (p < 0.05).

In order to remove the dependency of weight on height, BMI was used for comparison among the 13 villages (Fig. 5). In male BMI the five riverine/coastal villages did not significantly differ from one another, and had higher values than the inland/northern villages. When the elder men were excluded from the analysis, this pattern became more remarkable (right half of each sex in Fig. 5). For females, the coastal (Dorogori)

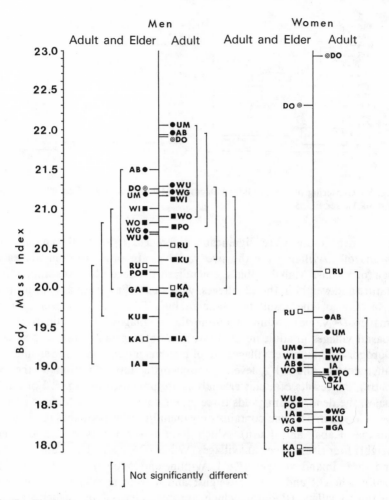

Fig. 5. The body mass index of the Gidra adults and elders of 13 villages.
□ Northern villages, Rual (RU) and Kapal (KA).
■ Inland villages, Iamega (IA), Wipim (WI), Podare (PO), Gamaeve (GA), Wonie (WO), and Kuru (KU).
● Riverine vllages, Ume (UM), Wuroi (WU), Woigi (WG), and Abam (AB).
◎ Coastal village, Dorogori (DO).

women had the highest BMI value, significantly different from the other 12 villagers, who differed little from one another.

In an attempt to simultaneously analyze stature and weight for 13 villages, multivariate analysis of variance (MANOVA) was performed. Since the resultant values of Wilks's λ were 0.7825 in men and 0.7923

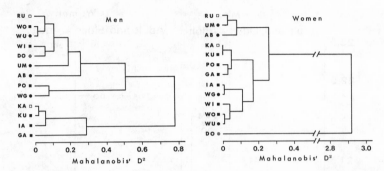

Fig. 6. Clustering of the 13 Gidra villages by Mahalanobis's D^2. For village names and marks, see Fig. 5.

in women, there were significant differences (p < 0.001) among the means of 13 villages for the two variables. In order to compare each pair of means, Mahalanobis's generalized distance (D^2) was applied to stature and weight in the 13 villages, with a test of each equality of means. Inspection of the resultant Mahalanobis's D^2 reveals a closer tie (i.e. smaller D^2 value) within inland/northern villages and within riverine/ coastal villages in adult males. In women, inter-group distances between Dorogori and all other villages are of high magnitude, showing significant differences at the 0.1 % level. For schematic representation of the D^2 matrix, the results of a cluster analysis are presented in Fig. 6. For adult males, the dendrogram yields three major complexes at a Mahalanobis's D^2 of less than 0.3: one contains only inland/northern villages (Gamaeve, Iamega, Kapal, and Kuru), which have lower values of stature and weight; four riverine/coastal villages (Wuroi, Dorogori, Ume, and Abam) and three inland villages (Rual, Wipim, and Wonie), which represent medium height and heavier weight; and a riverine village (Woigi) and an inland village (Podare), which are characterized by higher stature, form a cluster. For females, the coastal village (Dorogori) constructs a single cluster distantly separated from the other clusters (Fig. 6), with the highest position in both stature and weight.

Variation of Body Composition

Table 3 lists the anthropometric measurements and indices for four Gidra villages, namely, Rual (northern), Wonie (inland), Ume (riverine), and Dorogori (coastal), in 1981. The measurements were taken on almost all adults: 105 of 114 men (92%) and 125 of 138 women (91%). Many of them (98 men and 108 women) also had taken part in the 1980 survey, and hence they were measured twice, for stature and weight, on occasions one year apart. Changes in stature were minuscule in both sexes— that is, 0.23 ± 0.90 cm (0.13%) in men and 0.61 ± 1.08 cm (0.39%) in women, a level probably indicating unconscious intra-observer error. Weight changes were not great on average in either sex (men = 0.25 ± 1.74 kg, women = 0.33 ± 2.76 kg).

The subjects of the 1981 survey were on average about 1 cm taller in stature and 3 kg heavier in weight than all the Gidra adults measured in 1980. This may be due to the sampling of the four villages, which were selected from the four village groups but not sampled with probability for proportional size. The proportions of adult inhabitants of northern, inland, riverine, and coastal villages were 13, 50, 29, and 8% respectively, while those of measured subjects were 20, 20, 36, and 24%; thus the higher percentage of riverine/coastal villagers involved in the second survey might increase the grand mean of stature and weight.

Table 3 contains the results of ANOVA testing of the null hypothesis of equality of group means between the four villages. No significant difference among them was found in stature for both sexes or in calf circumference for men, while the other measurements significantly varied. The inland/northern males were lower in weight, chest, and arm circumferences, and two skinfolds (values lowest for the northern villagers) than the riverine/coastal males (values greatest for the coastal villagers). Means of these variables increased in the order of northern-inland-riverine-coastal (Rual-Wonie-Ume-Dorogori). For all the measurements, however, no significant difference by Duncan's multiple range test was found between the northern and inland villages, or between the riverine and coastal villages.

Means of female weight, two circumferences, and two skinfold thicknesses were significantly different among the four villages ($p < 0.001$). The coastal (Dorogori) villagers had the highest mean values in all the variables. Inland (Wonie) or northern (Rual) women represented the lowest means except for calf circumference, in comparison with the riverine or coastal women.

Either body composition, arm muscle size, or fat area had a similar pattern to weight, circumferences, and skinfold thicknesses. For men, variables assessing body fat, such as arm fat area, percent fat, and fat mass were higher in the riverine/coastal villages, whereas variation of muscle mass was relatively small. Fat-free mass was not significantly different among the four villages; the northern (Rual) village was significantly low in arm muscle size. For women, the coastal (Dorogori) subjects were by far the fattest and the most muscular, the other three showing small differences one from another.

Table 3. Means and Standard Deviations (SDs) of Anthropometric Measurements and Estimated Body Composition for the Gidra of Four Villages

Variable	Northern village: Rual (N = 21)		Inland village: Wonie (N = 22)		Riverine village: Ume (N = 38)		Coastal village: Dorogori (N = 34)		Four villages (N = 105)		ANOVA
	Mean[a]	SD	Mean[a]	SD	Mean[a]	SD	Mean[a]	SD	Mean	SD	F-value
Male											
Stature (cm)	165.31[A]	5.40	165.41[A]	4.94	166.72[A]	5.99	166.36[A]	5.18	166.08	5.44	0.44
Weight (kg)	54.43[B]	7.88	56.70[AB]	5.08	59.58[A]	8.16	60.58[A]	9.52	58.18	8.14	2.96*
Chest circ. (cm)	82.53[B]	4.19	83.73[AB]	3.09	84.92[AB]	4.67	86.25[A]	4.90	84.50	4.48	3.06*
Upper arm circ. (cm)	25.39[B]	2.12	26.55[AB]	1.57	27.15[A]	2.10	27.39[A]	2.45	26.73	2.19	4.22**
Calf circ. (cm)	33.35[A]	2.27	34.10[A]	1.93	34.29[A]	2.66	34.43[A]	3.30	34.10	2.61	0.77
Log triceps sf.[b]	149.52[B]	11.15[i]	150.86[B]	13.96	159.51[A]	10.38	166.46[A]	19.57	157.44	15.14[i]	7.35***
Log subscapular sf.[b]	184.30[B]	13.28[i]	184.64[B]	10.45	191.74[AB]	11.47	198.07[A]	21.66	190.33	15.35[i]	4.55**
Body mass index (kg/m^2)	19.83[B]	2.00	20.69[AB]	1.14	21.38[A]	2.29	21.81[A]	2.63	21.02	2.22	3.77*
Arm muscle circ. (cm)[c]	23.67[B]	1.94[i]	24.92[A]	1.53	25.32[A]	2.03	25.22[A]	2.30	24.90	2.05[i]	3.19*
Arm muscle area (cm^2)[d]	44.86[B]	7.24[i]	49.59[A]	6.03	51.32[A]	8.14	51.02[A]	9.09	49.69	8.07[i]	3.20*
Arm fat area (cm^2)[e]	6.21[C]	1.52[i]	6.71[BC]	1.69	7.69[B]	1.51	9.13[A]	3.50	7.54	2.38[i]	7.79***
Σ2 skinfolds (mm)[f]	14.09[B]	2.88[i]	14.21[B]	2.57	16.20[B]	3.06	19.63[A]	8.94	16.18	5.35[i]	6.06***
Body density (kg/m^3)[g]	1074.1[A]	6.0[i]	1073.8[A]	5.3	1069.6[AB]	5.5	1066.0[B]	11.4	1070.6	7.9[i]	6.07***
Per cent fat[h]	11.27[B]	2.37[i]	11.41[B]	2.11	12.95[AB]	2.21	14.56[A]	4.63	12.69	3.18[i]	6.10***
Fat-free mass (kg)	47.67[B]	6.24[i]	50.17[AB]	3.89	51.76[A]	6.42	51.44[A]	6.06	50.59	5.96[i]	2.30
Fat mass (kg)	6.20[B]	2.09[i]	6.54[B]	1.63	7.82[AB]	2.13	9.14[A]	4.36	7.56	2.90[i]	5.39**

Table3. continued

Female	(N=23)		(N=23)		(N=44)		(N=30)		(N=120)		
Stature (cm)	156.39[A]	4.89	155.22[A]	4.70	154.60[A]	4.76	156.51[A]	5.84	155.54	5.07	1.12
Weight (kg)	47.22[B]	6.66	44.56[B]	4.60	46.76[B]	8.45	55.25[A]	8.46	48.55	8.35	11.42***
Upper arm circ. (cm)	23.63[B]	1.65	23.63[B]	1.59	23.80[B]	2.48	26.24[A]	2.17	24.34	2.36	10.81***
Calf circ. (cm)	30.76[B]	1.93	30.32[B]	1.68	30.09[B]	2.67	32.29[A]	2.38	30.81	2.44	5.95***
Log triceps sf.[b]	175.63[BC]	21.58[k]	169.39[C]	22.17	182.88[B]	19.36	195.09[A]	19.52	182.07	22.07[l]	7.78***
Log subscapular sf.[b]	210.82[B]	19.88[k]	194.90[C]	19.00	211.46[B]	23.54	223.17[A]	20.32	211.10	23.03[l]	7.66***
Body mass index (kg/m²)	19.26[B]	2.20	18.53[B]	1.75	19.51[B]	3.07	22.56[A]	3.22	20.04	3.11	11.82***
Arm muscle circ. (cm)[c]	21.05[B]	1.22[k]	21.35[B]	1.21	20.90[B]	1.81	22.56[A]	1.34	21.44	1.63[l]	8.08***
Arm muscle area (cm²)[d]	35.39[B]	4.11[k]	36.37[B]	4.13	35.00[B]	6.05	40.65[A]	4.82	36.77	5.54[l]	8.14***
Arm fat area (cm²)[e]	9.24[B]	3.83[k]	8.26[B]	2.81	10.55[B]	4.86	14.50[A]	7.02	10.87	5.49[l]	7.99***
Σ2 skinfolds (mm)[f]	24.07[BC]	8.84[k]	18.84[C]	6.33	26.02[B]	11.46	32.47[A]	13.44	25.91	11.64[l]	7.16***
Body density (kg/m³)[f]	1048.1[B]	12.8[k]	1056.4[A]	11.7	1046.1[B]	14.5	1037.9[C]	13.5	1046.4	14.6[l]	8.29***
Per cent fat[h]	21.91[B]	5.31[k]	18.46[C]	4.78	22.74[B]	6.07	26.17[A]	5.78	22.63	6.14[l]	8.26***
Fat-free mass (kg)	36.77[B]	3.48[k]	36.23[B]	2.31	35.76[B]	4.64	40.40[A]	3.84	37.21	4.27[l]	9.43***
Fat mass (kg)	10.68[BC]	4.02[k]	8.33[C]	2.61	11.00[B]	4.75	14.85[A]	5.49	11.40	4.98[l]	9.68***

[a] Means with the same letter (A, B, etc) are not significantly different by Duncan's multiple range test ($p < 0.05$).
[b] $100 \times \log_{10}$ (skinfold thickness measured in 0.1 mm – 18).
[c] Arm muscle circumference = arm circumference – $\pi \times$ triceps skinfold.
[d] Arm muscle area = (arm circumference – $\pi \times$ triceps skinfold)^2/4π.
[e] Arm fat area = (arm circumference)^2/4π – arm muscle area.
[f] Σ2 skinfolds = triceps skinfold + subscapular skinfold.
[g] Body density was calculated from log Σ2 skinfolds, using the 20–29 year group equation of Durnin and Womersley (1974).
  Men: Body density (kg/m³) = (1.1525 – 0.00687 log Σ2 skinfolds) × 10³.
  Women: Body density (kg/m³) = (1.1582 – 0.0813 log Σ2 skinfolds) × 10³.
[h] Per cen fat = (4.570/body density – 4.142) × 100 (Brožek et al., 1963).
[i] Number of subjects = 19.
[j] Number of subjects = 103.
[k] Number of subjects = 21.
[l] Number of subjects = 118.
* p<0.05, ** p<0.01, *** p<0.001.

DISCUSSION

The Gidra form a biotic population, a reproductive unit occupying a particular space, whose members tend to share a gene pool. This characteristic was supported, at least tacitly, by the normal distribution of stature determined by the action of polygenes (Fisher, 1958; Harrison, 1977). In the case of data for a number of populations in the Enga-speaking people, such a single normal distribution was not found (Freedman and Macintosh, 1965). According to the studies on the resemblance of twins and familial correlations for stature (Carter and Marshall, 1978), a major part of the variation is largely due to additive polygenic inheritance, but environmental factors make some contribution to the variation. Indeed, the stature of the Gidra shows a reasonable amount of random variation without a special pattern related to ecological circumstances.

Stature difference between adults and elders can be attributed to two factors, the secular trend and true loss of height with age. During approximately the last hundred years in industrialized countries, there have been very striking tendencies towards acceleration in growth and increase in adult height, known as the "secular trend" (Tanner, 1962, 1968; Kimura, 1967; Ljung et al., 1974; Greulich, 1976; Meredith, 1976; Frisancho et al., 1977; van Wieringen, 1978). Some evidence for a secular trend in height is available from cross-sectional studies in Papua New Guinea: a significant fall of 2.3 cm in men and 5.0 cm in women from 20 to 60 years old in the Chimbu people (Maddocks and Rovin, 1965); an average reduction in height of 4.1 cm (2.6%) in males and 3.3 cm (2.2%) in females from the third to the seventh decade of age in the Enga-speaking people (Sinnett, 1975); a significant decline of 2 to 3 cm with age in the Lumi (Wark and Malcolm, 1969). On actual loss of height with age by individuals, Miall et al. (1967), in a longitudinal study of two Welsh communities, showed height decreases of 1.7 to 4.3 cm from 25 to 70 years old; Friedlaender (1975), who remeasured individuals in the North Solomons Province (Papua New Guinea), reported that those who were 25–29 years, 30–39 years, and over 40 years old at the time of first measurements (1938–39) were an average of 0.7 cm, 2.6 cm, and 2.8 cm shorter, respectively, than in the second measurments 28 years later (1967). It is judged that the decline of height in the Gidra could be accounted for as a true loss rather than a secular trend. The fact that height decline of females (1.56 cm) was greater than that of males (0.97 cm) may be associated with the women's frequent carriage of loads heavier than 40 kg or 50 kg; in contrast, men usually carry only bows and arrows in their hands (cf. Ohtsuka, 1983).

Decrease of weight in aging was noted in several Papua New Guinea populations for the Chimbu, Baiyer River, and Maprik peoples (Bailey, 1963), the Chimbu (Maddocks and Rovin, 1965), the Lumi (Wark and Malcolm, 1969), and the Bundi (Malcolm, 1970c). The main determinant in the decrease is considered to be age and bodily activity, combined with chronic malnutrition (Bailey, 1963). In other words, a negative energy balance in a marginal nutritional situation is accentuated by the heavy burden of physical labor (Malcolm, 1970c). Possible additional factors include chronic respiratory infection (Wark and Malcolm, 1969), although there is little support from the present data for the "maternal depletion syndrome" suggested by Jelliffe and Maddocks (1964), since the weight fall was more markedly observed in males than in females.

Among Papua New Guinea populations, the Gidra were the tallest and medium in weight, and hence their BMI was lowest, although they were shorter and lighter than the Europeans studied (Eveleth and Tanner, 1976). The mean percent fats, 12.69% for men and 22.63% for women, were comparable to those in developing countries (Norgan et al., 1982), but lower than those of Scottish residents (Durnin and Womersley, 1974). Adiposity was rare in the whole of the Gidra but was found in some Dorogori women. Both triceps skinfold thicknesses (men = 5.80 mm, women = 9.29 mm) and upper arm circumferences (men = 26.73 cm, women = 24.34 cm) fell at the 10th or 5th percentile of USA standard (Frisancho, 1981) and were lower than Jelliffe's standard (Jelliffe, 1966). The Gidra people are, however, muscular; upper arm muscle sizes (men = 24.90 cm, women = 21.44 cm) were comparable to Jelliffe's standard and calf circumferences (men = 34.10 cm, women = 30.81 cm) were similar to European values (Eveleth and Tanner, 1976). Thus the Gidra physique is characterized as medium-statured, lean, and muscular.

Values for weight, BMI, and Mahalanobis's generalized distance (except for stature per se) were greater for riverine/coastal males than for the inland/northern males, and this tendency was more marked in the younger generation. On the basis of comparison among the four selected villages, inland/northern (Rual and Wonie) males were lighter and leaner than the riverine/coastal (Ume and Dorogori) males, while the two groups were of nearly the same height and muscularity. Similar characteristics were recognized in females, but the female's data did not give such clear evidence: only the coastal (Dorogori) women were extremely tall, heavy, and fat as compared with the other 12 villages. Since relatively little study of morphological adaptations has been conducted on females, it is difficult to definitely explain such obscurity in the women's results. A possible factor may be the greater resistance of females to environmental influence which is alluded to as the "innate biological superiority of females" (Johnston, 1982: 362). The other conceivable reason is that inter-village migration due to residential moving at marriage was more frequent in females (Chapter 20 in this volume); 157 of 419 female subjects (37.5 %) left their birth villages while 70 of 332 men (21.1 %) changed their residence.

Apart from genetic determinants, food intake and physical activity are essential factors generating the intra-population variation, from the point of view of energy input and output. According to the result of our dietary survey (Chapter 9 in this volume), the nutrient intake varied among villages, reflecting differences in food consumption patterns which included different proportions of plant staples and animal protein sources, related to ecological (environmental and cultural) conditions. It is true that the intake of the "noble" nutrients (protein and fats) was higher in the riverine/coastal villages than in the inland/northern villages, but the energy intake of the former was inferior to that of the latter. The contradiction between the body physique and the energy intake may be relevant to the difference in physical activity between the two sub-groups. The inland/northern people travel on foot longer distances for hunting, sago-working, and food gathering and carry heavy loads such as carcasses, sago flour, garden crops, and firewood, whereas the riverine/coastal villagers can use canoes for transportation. This difference seems to be borne out by similar calf circumferences in the four villages in contrast to significant differences in upper arm circumferences.

(*Toshio Kawabe*)

Chapter 12

Infant Growth

In Papua New Guinea the bulk of knowledge on child growth rates has been accumulated in the last few decades. It is apparent from the studies so far made that the growth curve of children follows a similar pattern: rapid weight gain in the first few months of life—which is comparable to the Harvard Standard (Tanner *et al.*, 1966)—and a subsequent break and faltering in growth (Malcolm, 1979). Despite distinct genetic differences, such a tendency is found in many populations of the developing world. This implies that certain environmental factors are involved in determining the overall growth patterns of populations.

Environmental parameters that operate as the major determinants of growth rate include climate and altitude (Coon, 1954; Schreider, 1964), seasonality (Pagezy and Haupie, 1985; Crittenden and Baines, 1986), nutrition and pathogenic infections (Olson, 1975), and socio-economic factors such as urbanization, income level, and so on (Cook *et al.*, 1973; Eveleth and Tanner, 1976). Such psycho-behavioral variables as mother-infant bonding (Klaus *et al.*, 1981; Kobayashi and Brazelton, 1984) and cultural attitudes to childrearing (Jenkins *et al.*, 1984) have also been proposed as the significant factors. Among these, nutritional deprivation and frequent exposures to chronic infections are known to bring about growth failure (Jelliffe, 1955; Tanner, 1978; Martorell, 1980).

Since the growth rate of children in the developing world tends to drastically fluctuate during the early post-natal period, continuous observation for a shorter period of time—that is, the eco-sensitive period—is required (Suzuki *et al.*, 1984a). Incidence of disease and food intake should also be assessed at short time-intervals; e.g. every one or two weeks.

Using data from my own research on the Gidra, this paper illustrates early post-natal growth patterns, exemplifies their individual and short-term variations, and discusses the effects of environmental factors relating to child growth. Field work was conducted in Ume village in particular from July through December, 1981.

MATERIALS AND METHODS

In this study, the body weight of children aged 0–5 was measured every three weeks at 11:00 a.m. in Ume village. Thirty-nine children (23 males and 16 females) were sampled for an intensive follow-up study. Records of date of birth, birth weight, body weight (measured by the public health inspectors on March 10 and July 6, 1981) and any admissions to the Daru General Hospital were collected. Admissions to and discharges from the Aid Post at Ume village were also recorded from the daily records kept by the medical orderly at Ume, with additional information from my own observation and inquiries. For the calculation of age in this study, the decimal system was employed, following Tanner (1978).

The number of deciduous teeth was examined on August 3, 1981, for 14 children (nine males, five females), with the assistance of one of our research members, T.S., who is a medical doctor and was temporarily residing at Ume. Additional data from Wonie and Dorogori villages (seven males and two females) were collected by two other research members, R.O. and T.I., and used for the analysis.

Inquiries into childrearing practices (e.g. lactation, feeding, curing, and so on) were made of 36 women who had given birth. Frequency of lactation was investigated in eight lactating mothers over four successive days. Data for each mother were collected twice, in August and November 1981. Food consumption of three children (two males and one female) who have been completely weaned was examined for seven successive days.

In the late 1970s a small Aid Post was established in Ume; a native medical orderly works there, giving medical treatment. According to the village record book, kept by the local village official, clinic services and health inspections started in 1978, and polio vaccinations, triple antigen injections, and Sabin vaccinations have been given to children since that time.

The Ume villagers sometimes visit Daru, the provincial capital (a four-hr trip by motor-powered canoe) to sell local products (e.g. wild pig meat, taro, and banana), for admission to the Daru General Hospital, and for other reasons. A small village store sells imported goods, but supplies are irregular.

The impact of modernization is steadily increasing in the area between the inland villages and a relatively modernized coastal village, Dorogori, close to Daru. This trend is reflected in the diet, in levels of nutritional intake derived from imported food (e.g. rice, flour, canned fish, and sugar), which are higher in this intermediate area, Ume, than for people further inland who depend largely on sago and game animals, but lower than for the people in Dorogori, who depend more on imported food (Chapter 9 in this volume).

RESULTS AND DISCUSSION

General Growth Pattern of Children

The mean birth weight of Gidra children who were born in Daru General Hospital in the period between 1973 and 1981 and have remained in Ume, Dorogori, or Daru was 2.94 kg (N = 22, SD = 0.36) for males and 2.72 kg (N = 15, SD = 0.61) for females. No significant difference was found between the sexes (t = 1.22, p > 0.05). In comparison with the corresponding values reported from Papua New Guinea as a whole, which range widely, from 2.32 kg for the Asai and the Lumi (Wark and Malcolm 1969; Malcolm 1970a, 1970b) to 3.25 kg for the Simbu (formerly called Chimbu) (Barnes, 1963; Bailey, 1964a), the Gidra fall in the higher group.

Growth rates of 23 males and 16 females during the first three years

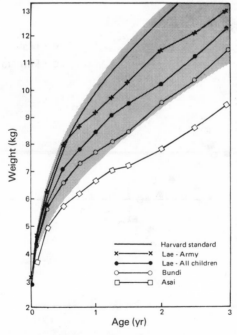

Fig. 1. Weight-for-age of the Gidra children (shown in the meshed area) compared with other Papua New Guinea populations and with Harvard Standard. Data source for Papua New Guineans: Malcolm, 1979.

are illustrated in Fig. 1 as a range between the minimum and maximum values. In this figure, the corresponding mean values of the four Papua New Guinea populations (Malcolm, 1979) and the Harvard Standard growth curve are shown. Despite the fact that the bulk of the Gidra children show lower weight-for-age than the Harvard Standard, they are comparable to or higher than other Papua New Guinea populations. As has been discussed in a previous report (Suzuki et al., 1984a), rapid growth gains during the first few months and successive declines are also noteworthy.

Growth velocities, shown as weight gains per year (kg/yr), among the same 39 children of Ume village were calculated between September 2 and December 21, 1981. Wide variations among different age groups are observed (Fig. 2). Growth retardation is marked among children after two years old.

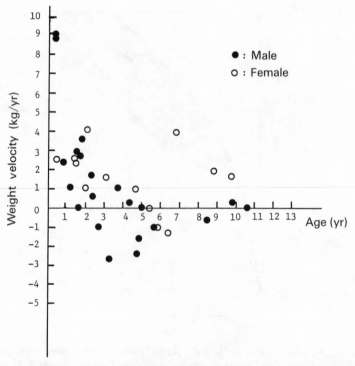

Fig. 2.   Weight velocity of Gidra children (September 2–December 21, 1981).

Lactation and Feeding Practices

In Gidra society there are taboos inhibiting pregnant and lactating mothers from taking certain kinds of food (cf. Wilson,1973). Thirty-four out of the 36 married women in Ume village who have experienced gestation stated that they used to, and do sometimes even now, observe food taboos. Items included are yams (13 cases), big fish (nine cases), snakes (seven cases), lizards (six cases), and so on. Two young women in their early 20s did not observe these customs.

It was not clearly recalled when mothers were released from the prohibition, but six of the 34 women stated that it corresponded to the time when a child started to crawl or to toddle; e.g. eight to ten months postpartum. However, my observation revealed that one mother ate wild pig meat when her child was about one month old. Another mother ate snake meat six months after delivery. It is unlikely that food taboos seriously affect the nutritional status of pregnant and lactating mothers.

Unlike bottle-feeding, which had come to prevail in urban sectors, but banned under national nutrition policy (Lambert and Basford, 1977; Lambert, 1980), breastfeeding is usual practice in the study area. Breastmilk is usually given to children on demand. Frequency of breastfeeding was investigated for eight lactating mothers during the four successive days. Inquiries were made twice in August and November. Children examined were two males and six females. Frequency of breastfeeding per day ranged from two to ten times (average for four days), and children of one year and over, took breastmilk no more than four times per day.

Among the Gidra lactation usually lasts as long as two years, and sometimes longer. In Papua New Guinea the lactating period varies from one group to another, and prolonged lactation appears to be common. For instance, in Lumi it is about 3.5 years (McKay, 1960), about 4 years in Simbu, Bundi, and Asai (Venkatachalam, 1962; Bailey, 1964a; Malcolm, 1970a), and 4.5 years in Kyaka (Becroft, 1967a, 1967b). Gidra's case shows that it is shorter than other populations.

In Gidra society the introduction of supplementary food commences quite early. Mothers start to give ripe banana of specific varieties to infants when they reached one or two months old. Ripe banana is first premasticated by mothers and then given to the babies. In one observed example, three bananas (each weighing 50 g) were first given to one newborn baby 20 days postpartum in December, 1981 (Fig. 3). A few other solids (taro, pumpkin, sweet potato, and so on) are subsequently added to the daily menu of the children. The first item of animal food given to babies is usually freshwater prawn, at around six months old. Thus, kinds

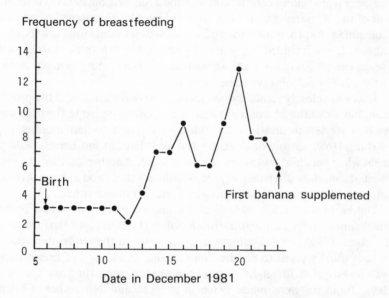

Fig. 3. Frequency of breastfeeding and the introduction of supplementary food of a male baby (December 6–23, 1981).

of supplementary foods increase as a child grows. According to a report on the timing of the introduction of supplementary food in Papua New Guinea (Clarke, 1978), in 33 out of 55 populations studied, people give initial supplementary food to children between six and 12 months postpartum (normal stage); three populations do so within one month (very early stage), eight between one and six months (early stage), nine between one and two years (late stage), and two over two years (very late stage). This suggests that the Gidra's practice falls into the early stage category.

Growth and Food Intake of Children

Individual variations in food intake and feeding methods were investigated for 16 children (11 males and five females) less than three years old. Numbers of food items consumed by individual children differ by age and by kind of food; numbers of animal foods consumed are greater than those of vegetables, and the former is more varied in kind than the latter among individual children ($F = 12.47$, $p < 0.001$).

The feeding method also changes in line with child growth. It is not until 14 months of age that children start to eat solid food without the aid of maternal premastication, though there are individual variations (Fig. 4).

Figure 4 also shows the number of deciduous teeth for 23 children (17 males and six females); the number is similar to that in other groups in Papua New Guinea (Bailey, 1964b; Barker, 1965; Malcolm, 1970c). It should be noted that the switch of feeding methods roughly corresponds to the time when children have eight to 12 deciduous teeth.

Among 18 children investigated, the three who had completed lactation were between one and two years old. Based on the nutritional values of Gidra food (Chapter 7 in this volume), energy and nutrient intake of these three children was measured for seven successive days. Daily energy intake per unit body weight varies from 70 to 110 kcal/kg, and daily

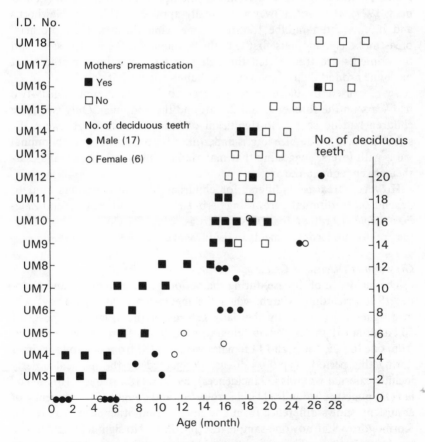

Fig. 4. Feeding method and number of deciduous teeth among Gidra children.

protein intake per unit body weight ranges between 0.7 and 1.8 g/kg. These values are lower than FAO/WHO recommended daily intakes for energy and protein (FAO/WHO, 1973). Despite a small sample, protein-energy malnutration of the children may be one of the important factors in causing growth retardation (Malcolm, 1979; Burman, 1982).

Disease and Healing for Children

In their humid tropical environment, Gidra children suffer from various kinds of infections and diseases. General accounts of disease and healing among the Gidra have already been reported by Suzuki (1985). Traditional healing is characterized by the use of such medicinal plants as *Acorus calamus* and *Zingiber officinale* and by a bleeding technique using a bivalve shell razor (*Gelonia* sp.) to eliminate "bad blood" from the body. Bleeding is also a common practice in lowland Papua (Landtman, 1927). It is not, however, normally applied to children. Newborns and infants often imbibe traditional medicine through their mothers' breastmilk; e.g. mothers first take the medicine for the babies' sake, and the contents are transmitted through nursing. Sometimes mothers put medicine around their nipples so their babies will suck it.

Scars caused by bleeding were observed in 37 children. Thirty-two had scars on the forehead, and 28, around the abdomen. Only two male children had no scars. No significant difference was found between the sexes (p > 0.05). The bleeding is undertaken for headaches and abdominal pains with fevers, suggesting that malaria and intestinal disease and diarrhea are endemic in the area.

However, treatment differs; for childrens' diarrhea mothers tend to depend on traditional cures using herbs such as *Acorus calamus* and *Zingiber officinale*, rather than western drugs, while for children's high fever they prefer modern treatments to traditional cures (Akimichi, 1985).

Growth and Disease of Children

The incidence of disease during the period between January and October 1981 (excluding March, when the medical orderly was away) was investigated, using mainly the daily records of the village aid post. Of 50 children (31 males and 19 females) aged 0–12, 19 males were treated from one to nine times and 14 females were treated from one to four times during this period. Types of disease designated by the medical orderly include external wounds (29 incidences), measles (18), scabies (16), malaria (11), puncture wounds (11), diarrhea (9), and coughs (7). Frequency of admission shows difference from one month to another according to the Kormogorov-Smirnov one-sample test (p < 0.01). No significant difference was found between the sexes with regard to susceptibility to diseases.

Of 50 children eight had malaria in June, and 18 had measles in July and August. The officer in the Malaria Section of Daru General Hospial has noted that cases of malaria in the Western Province increase just before the onset of the wet season (Siwani, 1981); this parallels the general seasonal pattern of malaria (Bray, 1979) and the Gidra's recognition of the malarial season (Suzuki, 1985). An unpublished epidemiological report for 1980–1981 (Division of Health, Malaria Service of the Daru General Hospital) also demonstrates that *Plasmodium vivax* and *P. falciparum,* the agents of malaria, are dominant in blood samples (Daru General Hospital, 1981). However, in the present data from Ume, the highest incidence of malaria occurred at the end of the wet season or the beginning of the dry season.

The age of children infected with measles ranges from four months to five years; among the 11 children less than 18 months old, only four did not suffer from measles. It was ascertained that of the 18 children who were infected with measles, ten were five pairs of co-residential siblings.

Because the people were in the bush camp during November and December to procure foods (sago and animal meat) for funeral and festive use, precise data were unavailable for those months; however, no less than 13 children (out of 50) were recorded as having severe colds, apparently as a result of exposure to rain and cold at the camp site.

Given the prevalence of various diseases at different times of the year, growth velocities in weight (kg/yr) of children suffering from infections between March 10 and October 18, 1981, were lower than those of uninfected children (Table 1).

Those who were infected by disease and treated at the aid post gained less weight than the others ($t = 2.76$, $p < 0.05$). Despite the small number of subjects, this difference was particularly noteworthy in children less than six months of age.

Individual Follow-up of Weight Change

To assess short-term fluctuation in child growth, a continuous examina-

Table 1. Weight Velocities of Infected and Non-Infected Children

	N	Mean ± SD (kg/yr)
Infected children	24	1.70 ± 1.03
Non-infected children	13	3.68 ± 2.37
t-value		2.76*[a]

* $p < 0.05$; [a] by Welch's method ($F = 5.50$, $p < 0.01$).

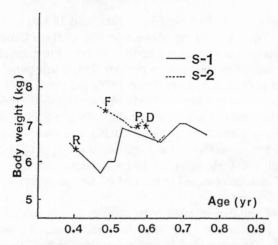

Fig. 5. Two examples of continuous weight-for-age observation of Gidra children and their history of illness, based on my measurement. R: measles, F: high fever, D: diarrhea, P: suppuration.

tion of weight change in individuals was made between August and December 1981. Trends in weight change of six children as examples, designated S-1 to S-6, are described in Figs. 5 and 6 to demonstrate the differential effects of infectious diseases on the growth rate.

S-1 (female) was infected with measles in early August, 1981 and experienced subsequent weight loss; in September her weight returned to normal, but no marked increment was observed after October. S-2 (male) was almost one kg heavier than S-1 at the beginning of the observation in August 1981, but then he contracted a high fever (perhaps due to malarial infection), and the subsequent weight loss due to suppuration and diarrhea caused his death (Fig. 5).

S-3 (male) was infected with malaria in early June 1981, with suppuration in early July, and with measles on July 21, when he was almost two years of age. Despite the weight loss in this period, marked weight gains were observed after recovery, as late as early September. S-4 (female) was infected with measles in early August and later had a high fever and a rash, but she did not lose weight during the early post-natal period (Fig. 6).

S-5 (female) and S-6 (male) are twins; both were adopted at the age of 11 months and two weeks, but by different families. They were not breastfed after that time. A significant difference in weight change was observed between August 12 and December 29 as a result of covariance analysis ($F = 119.5$, $p < 0.0001$); the difference in slopes was not signifi-

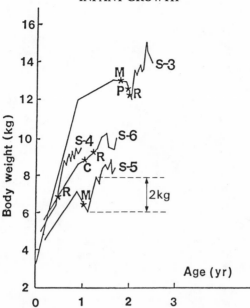

Fig. 6. Four examples of weight-for-age observation of Gidra children and their history of illness, based on records and my measurement. M: malaria; R: measles; C: cough; P: suppuration.

cant, and the difference in elevations was large, between 1.1 kg and 2.4 kg, during my observation period. As is clearly shown in Fig. 6, S-5 gained 2 kg during June and July despite the fact that she lost weight due to weaning and the subsequent malarial infection in June; in S-6 no weight loss was observed despite the fact that he was infected with cold and contracted measles.

In summary, these examples indicate that such diseases as malaria, measles, and infantile diarrhea had differential effects on the growth rates of children, ranging from apparent growth failure to steady weight gain. Catch-up growth after recovery from illness (Prader *et al.*, 1963) was observed in some cases. Thus, the present data on individual variations in the growth pattern demonstrate not only differentiated effects of disease on growth but also a need for short-term observations.

(Tomoya Akimichi)

Chapter 13
Growth of Children and Adolescents

The study of growth and development using historically established methods is dependent on birth records. The apparent lack of such records, combined with limited information on age, has caused the scarcity of works on nonliterate peoples. Growth studies in Papua New Guinea have been devoted to subjects with accurate age records based on mission baptismal registers or other sources (Wark and Malcolm, 1969; Malcolm, 1969a, 1970c, 1970d; Heath and Carter, 1971; Harvey, 1974). Elsewhere, age has been estimated from dental eruptions (Malcolm, 1969b) or combinations of birth registrations, family relationships, age at known events in the past, and appearance (Walsh et al., 1966; Wark and Malcolm, 1969; Sinnett and Whyte, 1973), or uncritically determined (Champness et al., 1963; Wolstenholme and Walsh, 1967). In spite of all precautions, errors are inevitable in these age estimations. It is therefore necessary to develop a method for analyzing the data with incomplete age records. In the present analysis, using the new method for fitting growth curves, two logistic functions are introduced for describing prepubertal and adolescent growth curves, and switching the former to the latter, based on annual increment data from two measurements.

In the 1980 survey, I measured statures (only for subjects who were able to stand upright) and weights of 840 children (432 boys and 408 girls), which corresponded to 82.5% of the total child population (1018); the proportion of the measured to the total children of each village ranged from 71.2% to 93.9%. Most of the children present in the survey period participated in the anthropometric measurements; the exceptions were babies and some older children who feared the Japanese investigators. In 1981, approximately one year after the initial survey, 420 boys and 379 girls were examined. Statures and weights were measured again, and five measurements—chest circumference (only for boys), upper arm circumference, calf circumference, triceps skinfold, and subscapular skinfold—were added for adolescents. Infants with only weight values, who were unable to stand with their heels together by themselves, and the physically handicapped are excluded from the present analysis. There is no evidence of severe malnutrition associated with edema, according to our observation but not according to a diagnosis by a medical specialist. Thus, partitions of 383 boys and 353 girls in 1980, and of 339 boys and 306 girls in 1981, were submitted to analysis.

Although 319 boys and 309 girls were examined twice, annual growth analysis was conducted only for subjects for whom both height and weight were known—273 boys and 262 girls. Since intervals between the two measurements were one year plus or minus two months at the most (401 out of 537 subjects: 75% within the limits of 1 ± 0.083 yr; 97% within 1 ± 0.167 yr), seasonal effects were minimal. To obtain interval differences among subjects, a 1-yr increment is calculated for each individual; the difference between 1980 and 1981 values is divided by the interval in decimal years between the two measurements (for decimal years, see Tanner, 1978: 172).

Only 42 of 840 subjects had birth date records by Daru Hospital, and the birth years of a further 17 children were determined by reliable information on age. In spite of our best efforts at age estimations based on birth order information passed on by the residents of each village, family relationships, and age at known events in the past (especially for the four villages intensively investigated: Rual, Wonie, Ume, and Dorogori), the ages of only 82 subjects were estimated with satisfactory accuracy (to within six months). Therefore the number of subjects with data available for both stature and age is 141 (76 boys and 65 girls).

GROWTH CURVE FOR THE GIDRA

Mathematical methods to fit longitudinal growth data have been used for description and interpretation of physical growth. Curve fitting of relatively simple mathematical models to growth data makes it easy to compare individuals by reducing large amounts of growth data to a small number of parameters for individual children. More importantly, it will enable us to compare growth patterns between populations or sub-populations, using these parameters.

There is a considerable degree of independence between growth before and at adolescence, probably because genes controlling the magnitude and timing of the adolescent growth spurt may produce no effect until the moment when the secretion of androgenic hormones begins (Tanner, 1977). Indeed, the complexity of human growth has obliged us to divide the curve into cycles (or periods) and to treat each of them separately rather than to treat the whole. For the preadolescent period, the mathematical models widely used for describing growth curves are the Jenss-Bayley model, i.e. $y = a + bt - \exp(c + dt)$, on length and weight

(Jenss and Bayley, 1937; Deming and Washburn, 1963) and the Count model ($y = a + bt + c\log t$) on skull dimensions (Count, 1942), height (Count, 1943; Israelson, 1960), and other measurements (Tanner et al., 1956). The logistic and Gompertz functions have been used during the adolescent period; the logistic for height by Marubini et al. (1971) and for other measurements by Marubini et al. (1972) and Tanner et al. (1976); the Gompertz for length by Deming (1957) and for weight by Laird (1967). Marubini et al. (1972) affirmed that the logistic function, in nearly all aspects, fit the data better than the Gompertz at adolescence. Recently, models which could describe growth curves from an early age to maturity have been proposed, such as the double logistic function (Bock et al., 1973; Rarick et al., 1975; Thissen et al., 1976; El Lozy, 1978) and the Preece-Baines model (Preece and Baines, 1978; Hauspie et al., 1980). None of these methods, however, was applicable to a growth study of population with incomplete age records, even if they were convincingly applied to longitudinal growth data.

In this study, a new method for fitting growth curves is used to meet the following two requirements: it should be applicable to data with little information on age; and it should treat a wide range of the growth process from early age to maturity. Two logistic models are introduced, starting from the most general logistic function, to describe the pre-pubertal and adolescent growth curves, switching from the former to the latter. The essential procedure of estimating parameters of a logistic function was exploited by Masuyama (1980, 1981, 1982), though he did not aim at meeting the requirement of application to data without complete birth records.

Kawabe (1986b) gave a complete mathematical description of the logistic growth model and applied it to the Gidra data without complete birth records. The resultant growth curves of stature are represented as two logistic functions smoothly switching from the prepubertal to the adolescent curves, for boys,

$$\text{Stature (cm)} = \begin{cases} \dfrac{167.122}{1 + \exp\{-0.16734\,(t - 2.3649)\}} & \text{if } t \le 10.3\text{yr} \\[2ex] 165.44 - \dfrac{43.550}{1 + \exp\{0.59061\,(t - 12.3021)\}} & \text{if } t > 10.3\text{yr} \end{cases}$$

and for girls,

$$\text{Stature (cm)} = \begin{cases} \dfrac{168.483}{1 + \exp\{-0.16582\,(t - 2.5165)\}} & \text{if } t \le 8.7\text{yr} \\[2ex] 154.95 - \dfrac{43.666}{1 + \exp\{0.59050\,(t - 10.2048)\}} & \text{if } t > 8.7\text{yr} \end{cases}$$

where t is age in year.

The goodness of fit of the logistic curves is examined in the comparison of the estimated growth curves with the average values by using reported longitudinal data (Kawabe, 1988). The result will be thoroughly described in a forthcoming paper.

COMPARISON OF GROWTH CURVES

The height growth curves of Gidra male and female children, estimated by using the logistic functions, are graphically compared in Fig. 1, with those of Papua New Guinea and some developed countries: the standards of USA (Hamil *et al.*, 1979), British (Tanner *et al.*, 1966), and Japanese (Ministry of Education, Science and Culture of Japan, 1985) children. The growth curves of the Gidra children are substantially below those of children in the developed countries, showing a parallel relationship. Recently, growth and development of Papua New Guineans has received special attention because of their remarkably slow growth rates, particularly among the highlanders, in comparison with other populations (Malcolm 1969a, 1970a; Wark and Malcolm, 1969). The Gidra boys and girls are taller than the other Papua New Guinea children, especially in

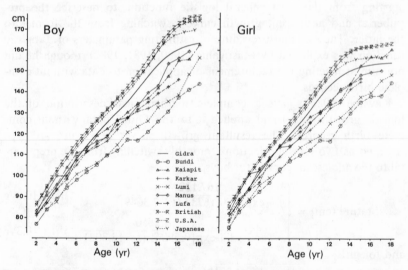

Fig. 1. Estimated growth curve of stature of the Gidra children compared with six groups of Papua New Guinea children and three standards in developed countries: Bundi (Malcolm, 1970c); Kaiapit (Malcolm, 1969a); Karkar (Harvey, 1974); Lumi (Wark and Malcolm, 1969); Manus (Heath and Carter, 1971); Lufa (Harvey, 1974); British (Tanner *et al.*, 1966); USA (Hamil *et al.*, 1979); Japanese (Ministry of Education, Science and Culture of Japan, 1985).

the adolescent period. The Kaiapit children overtake the Gidra at the end of adolescence, and indeed the Kaiapit adult is the tallest among the seven Papua New Guinea peoples (for adult stature, see Chapter 11 in this volume). The growth of the Gidra is faster than that of the Kaiapit, but their adult height is less than that of the Kaiapit for both sexes. This result is the same as that for the Manus (Heath and Carter, 1971) and does not support Malcolm's (1970d) suggestion that adult height may be proportionally related to child growth rate in traditional Papua New Guinea societies.

DIFFERENCE WITHIN THE GIDRA POPULATION

In order to compare growth of the Gidra children between inland/ northern and riverine/coastal areas, parameters of the logistic functions were computed for these two groups. The results are illustrated in Fig. 2. Examination of the curves reveals that growth rates for inland/northern children are less than those for riverine/coastal children, especially in boys. The riverine/coastal boys exhibit more rapid growth in the prepubertal period than do the inland/northern boys. The former's stature exceeds the latter's, and the difference increases in the early stage of the

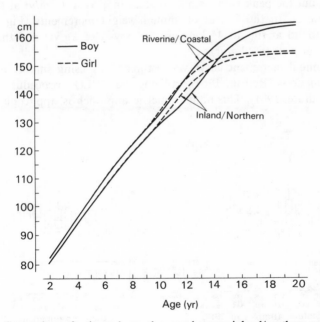

Fig. 2. Comparison of estimated growth curves between inland/northern and riverine/ coastal groups.

adolescent growth spurt; the latter's stature approaches that of the former at the end of the adolescent period. On the other hand, for the prepubertal growth of girls, the graph shows that the height of riverine/coastal girls is slightly superior to that of inland/northern girls. The two consequent curves are parallel in adolescence, showing higher stature in the riverine/ coastal girls. The girls' stature difference between the two groups diminishes at the end of adolescent period, as it does among boys.

ANNUAL INCREMENTS OF STATURE AND WEIGHT

Annual increments of stature and weight are widely scattered. Mean and SD values are calculated for each class grouped in 5 cm increments of stature in 1980 and shown with median, maximum, and minimum values in Figs. 3 and 4. In boys' stature, the mean annual increment is 7 cm/yr at the 90–95 cm level and gradually declines to 4.7 cm/yr (130–135 cm level) with small fluctuations. It rapidly increases to the maximum value of 8.2 cm/yr at the 145–150 cm level, with large SDs which may be due to individual variations at the time of the adolescent growth spurt. After the peak is attained, the mean increment sharply diminishes. The changing pattern of girls' annual stature increments is similar to that of boys, but the peak value is not so great in girls: 6.1 cm/yr at the 135–140 cm level. In the figures of annual weight increments (Fig. 4), the peak is found at the 40–45 kg class for boys (6.5 kg/yr) and the 35–40 kg class for girls (6.2 kg/yr).

The annual increment curve was estimated by using smoothing cubic spline functions (Reinsch, 1967), utilizing the GPLOT procedure of SAS (SAS Institute, 1981). This nonparametric approach is applicable for the

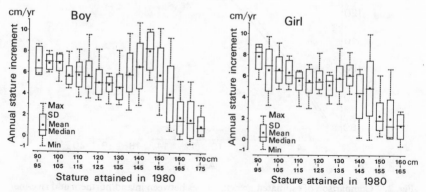

Fig. 3. Annual increment of stature plotted against stature attained in 1980.

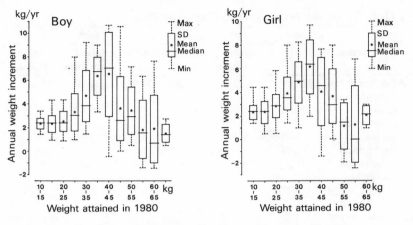

Fig. 4. Annual increment of weight plotted against weight attained in 1980.

increment data, since no fixed parametric function needs to be postulated *a priori.* Smoothing constant of the parameters of SAS was determined as L = SM40 (SAS Institute, 1981) by the trial-and-error approach. The resulting curves on the data for 273 male and 262 female children are shown in Fig. 5. Parameters such as peak height velocity and height at peak velocity are obtained directly from the curves estimated by the spline function and given in Table 1. The beginning of the adolescent spurt, called "take-off," is defined graphically as the point of minimum velocity, and peak velocity signifies the maximum velocity during the growth spurt;

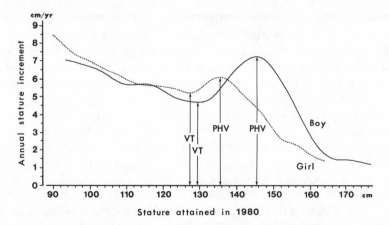

Fig. 5. Annual stature increment curves estimated by smoothing cubic spline functions, showing peak height velocity (PHV) and velocity at take-off (VT).

Table 1. Height and Velocity at Take-off and Peak Point

Variable	Boy			Girl		
	All	Inland/ northern	Riverine/ coastal	All	Inland/ northern	Riverine/ coastal
Adult height (cm)	165.44	165.04	166.05	154.95	154.69	155.48
Height at take-off (cm)	129.5	131.0	130.6	127.4	126.8	126.7
Velocity at take-off (cm/yr)	4.7	4.4	5.2	5.2	5.0	5.8
Height at peak velocity (cm)	145.5	146.0	146.3	135.6	135.2	136.3
Peak height velocity (cm/yr)	7.2	7.4	7.3	6.1	5.9	6.8
Height increase, take-off to PHV[a] (cm)	16.0	15.0	15.7	8.2	8.4	9.6
Velocity increase, take-off to PHV (cm/yr)	2.5	3.0	2.1	0.9	0.9	1.0
Percentage adult height at take-off	78.3	79.4	78.7	82.2	82.0	81.5
Percentage adult height at PHV	87.9	88.5	88.1	87.5	87.4	87.7

[a] PHV: peak height velocity.

these are marked by arrows in the diagrams. The adolescent spurt starts in girls at a height of 127.4 cm, which is 82.2% of adult stature (154.95 cm), and velocity at take-off (VT) is 5.2 cm/yr. In boys the spurt begins 2.1 cm higher than in girls, at 129.5 cm, 78.5% of adult value (165.44 cm); VT = 4.7 cm/yr. Difference in height at peak velocity between girls (135.6 cm) and boys (145.5 cm) is 9.9 cm, but percentage of adult height is about 88% in both sexes. Peak height velocity (PHV) is 6.1 cm/yr in girls and 7.2 cm/yr in boys. Velocity increase during the growth spurt, or the difference between PHV and VT, in boys is 2.5 cm/yr; in girls it is 0.9 cm/yr, less than half of boys' value.

There is another small peak at the height of 115 cm in boys, and a diminution of deceleration at about 120 cm in girls, which suggests the occurrence of a mid-growth spurt, occurring several years before the adolescent spurt, as labelled by Tanner (1947). Tanner and Cameron (1980) reported that London boys had a diminution of deceleration, or a relative spurt, in height from age six to seven years in single-year velocity data, and Berkey et al. (1983), using a variable knot cubic spline method, found a mid-growth spurt in height for 17 of 67 Boston boys but for no girls. Using kernel estimates, Gasser et al. (1985) confirmed that the small mid-growth spurt is a constant phenomenon in Zürich children of both sexes.

The estimated smooth curves in weight are not so well-fitted since the weight at peak velocity was too small, especially in boys, in comparison with the results of the 5 kg distance grouping (Fig. 4). This might be due to the existence of subjects who exhibit small annual increments (near zero) in the weight class of 35–45 kg.

The annual stature increment curve for each of the inland/northern and riverine/coastal groups is estimated by the same smoothing method; the resultant parameters are listed in Table 1. Peak height velocity of the inland/northern boys (7.4 cm/yr) is similar to that of the riverine/coastal boys (7.3 cm/yr) at the height of around 146 cm, while the velocity at take-off of the former (4.4 cm/yr) is lower than that of the latter (5.2 cm/yr). Velocity increase during the adolescent spurt is greater in the inland/northern group (3.0 cm/yr) than in the riverine/coastal group (2.1 cm/yr). In other words, the riverine/coastal boys maintain a higher velocity than the inland/northern boys who reach the same velocity at the peak. Riverine/coastal girls were superior to the inland/northern girls by about 1 cm/yr during most of the study period.

RELATIVE GROWTH

Form or shape during growth is a function of absolute size rather than of absolute age (Reeve and Huxley, 1945). Relative growth or allometry allows us to ignore the time relations of growth by relating the sizes of the different parts of the body to total size, regardless of age (Shimizu, 1959). This implies that allometry can be used to analyze data with no or little information on age, like that for the Gidra population.

Relationships between sizes and shapes of organs and organisms have long been of interest to biologists, and various scholars have pointed out that certain organs show a marked progressive change in relative size with an increase in absolute size (Reeve and Huxley, 1945). Thompson (1942) stressed that all organic forms are the result of differential growth. Huxley's work (1932), a classic in this field, placed the subjects on a firm quantitative basis and pointed the way to a deeper understanding of many phenomena connected with organic growth. Since then, the study of the relative growth of parts of organisms has become more widespread and has demonstrated that allometry occurs almost universally in animals (for an excellent account of this topic, with a list of over 200 references, see Gould, 1966).

In the present analysis, allometry (or relative growth) is used in the meaning of the very general definition given by Gould (1966), the differences in proportions correlated with changes in absolute magnitude of the total organism or of the specific parts under consideration. Most studies have been concerned with the so-called equation of simple allometry (Huxley and Teissier, 1936), which is expressed either as

$$y = bx^{\alpha},$$

or in the logarithmic form as

$$\log y = \log b + \alpha \log x.$$

If the logarithms of two dimensions x and y obeying the law are plotted against one another, the points lie along a straight line. Three ways of plotting were, however, adopted in the present analysis, in conformity with Hiernaux (1964): first, values without any transformation of anthropometric measurements against stature; second, log values against stature (semi-logarithmic plot); and third, log values against log stature (simple allometry).

Regression lines were calculated by the usual least squares method (y on x) when rectilinearity was assumed. The line has the general formula $y = a + bx$ in which x is stature (cm) or its log value, y the other measurements with or without logarithmic transformation, a the intercept of the line, and b its slope. Comparison between groups is reduced to comparing the 'a's and 'b's. By this method, we can eliminate two factors, duration of growth and final size, which may vary between the groups compared (Hiernaux, 1968).

Allometry in the Whole Gidra Child Population

The changes in relationship between weight and stature (in the measured values and the transformed values) during growth are shown on diagrams, using the data for 339 boys and 306 girls (Fig. 6). The figures illustrate the non-linear relationship, indicating some kind of exponential relation between stature and weight, similar to those of African children reported by Hiernaux (1964) and of Papua New Guineans described by Malcolm (1970c). The log weight/stature relationship is illustrated at the middle of Fig. 6. Some kind of linear relationship is suggested, since there appears to be no (or little) indication of curvature. Tables 2 and 3 list the parameters of the regression lines and the correlation coefficients, which are remarkably high (0.984 for boys and 0.973 for girls) and indicate high degrees of fitting. Log weight/log stature diagrams (at the bottom of Fig. 6) appear to be less strictly linear than on the log weight/stature ones, since the correlation coefficients of the former (0.976 for boys and 0.962 for girls) are lower than those of the latter.

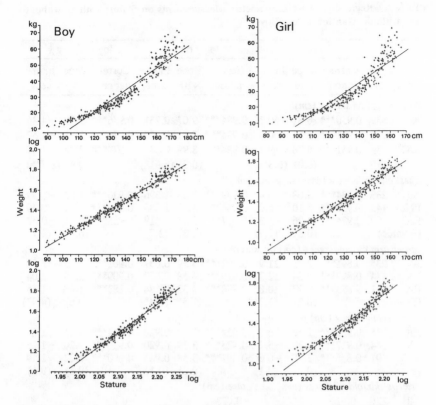

Fig. 6. Weight for stature relationship: weight/stature (top); log weight/stature (middle); log weight/log stature (bottom).

Regressions of the chest, upper arm, and calf circumferences and two skinfold thicknesses are represented in Tables 2 and 3. All but the triceps skinfold of boys prove to have significant linear relationships, representing a better fit to the regression line in the log values/stature case than in the non-transformed and log-log ones, although the linearities are less than those of the weight in both sexes. There is no greater tendency toward augmentation of boys' triceps skinfold with increases in stature than with age, but there is wider variation.

When the logarithm of the weight or dimension of parts of an organism is plotted against stature, i.e. on log-log scale, the plotted points often appear to fall on or near a straight line (Huxley, 1932). This would imply an exponential relationship between the variables of the form $y = bx^{\alpha}$, which is usually known as simple allometry (Huxley and Teissier, 1936).

Table 2. Regression of Anthropometric Measurements on Stature with or without Logarithmic Transformation: Boys

Group	N	y on x Corre-lation	y on x Slope ×10	y on x Inter-cept	log y on x Corre-lation	log y on x Slope ×10³	log y on x Inter-cept	log y on log x Corre-lation	log y on log x Slope	log y on log x Inter-cept
Weight (kg) on stature (cm)										
All	339	0.950***	6.81	−60.3	0.984***	9.07	0.231	0.976***	2.79	−4.46
I/N[a]	205	0.951***	6.82	−60.9	0.984***	9.11	0.220	0.976***	2.81	−4.52
R/C[b]	134	0.951***	6.78	−59.1	0.984***	8.99	0.247	0.978***	2.76	−4.39
(F-value)[c]			(0.0)	(6.3*)		(0.5)	(6.9**)		(0.6)	(8.7**)
Chest circumference (cm) on stature (cm)										
All	233	0.934***	5.08	−3.4	0.939***	3.08	1.396	0.936***	1.03	−0.39
I/N	143	0.937***	5.10	−4.0	0.944***	3.13	1.388	0.940***	1.04	−0.40
R/C	90	0.924***	4.96	−1.3	0.927***	2.95	1.419	0.925***	1.01	−0.33
(F-value)			(0.3)	(1.7)		(1.3)	(2.7)		(0.3)	(2.2)
Upper arm circumference (cm) on stature (cm)										
All	235	0.891***	2.23	−11.9	0.907***	4.59	0.637	0.901***	1.53	−2.00
I/N	144	0.899***	2.24	−12.2	0.916***	4.68	0.620	0.909***	1.55	−2.04
R/C	91	0.873***	2.17	−10.8	0.888***	4.32	0.684	0.883***	1.46	−1.85
(F-value)			(0.2)	(3.3)		(1.5)	(5.4*)		(0.6)	(4.9*)
Calf circumference (cm) on stature (cm)										
All	235	0.906***	2.34	−6.0	0.914***	3.59	0.920	0.914***	1.21	−1.16
I/N	144	0.917***	2.32	−5.7	0.925***	3.59	0.920	0.926***	1.20	−1.15
R/C	91	0.886***	2.38	−6.6	0.891***	3.58	0.923	0.890***	1.22	−1.19
(F-value)			(0.2)	(0.1)		(0.0)	(0.1)		(0.1)	(0.1)
Triceps skinfold thickness (mm) on stature (cm)										
All	226	0.013			−0.023			−0.019		
I/N	135	−0.006			−0.045			−0.043		
R/C	91	0.005			−0.029			−0.024		
Subscapular skinfold thickness (mm) on stature (cm)										
All	226	0.571***	0.84	−4.8	0.622***	4.75	0.162	0.618***	1.59	−2.58
I/N	135	0.614***	0.86	−5.3	0.655***	5.01	0.115	0.649***	1.65	−2.73
R/C	91	0.493***	0.78	−3.5	0.550***	4.12	0.273	0.549***	1.40	−2.15
(F-value)			(0.3)	(2.3)		(1.2)	(3.7)		(0.8)	(3.6)

[a] Inland/northern group.
[b] Riverine/coastal group.
[c] Equality of slope and intercept by ANACOVA.
* $p < 0.05$; ** $p < 0.01$; *** $p < 0.001$.

Table 3. Regression of Anthropometric Measurements on Stature with or without Logarithmic Transformation: Girls

Group	N	y on x			log y on x			log y on log x		
		Corre-lation	Slope ×10	Inter-cept	Corre-lation	Slope ×10³	Inter-cept	Corre-lation	Slope	Inter-cept
Weight (kg) on stature (cm)										
All	306	0.923***	6.66	−58.1	0.973***	9.95	0.117	0.962***	2.89	−4.69
I/N[a]	217	0.932***	6.50	−56.4	0.977***	9.90	0.119	0.967***	2.88	−4.67
R/C[b]	89	0.912***	7.12	−62.9	0.967***	10.14	0.105	0.954***	2.94	−4.78
(F-value)[c]			(3.3)	(6.5*)		(0.7)	(9.9**)		(0.4)	(6.8**)
Upper arm circumference (cm) on stature (cm)										
All	197	0.842***	2.22	−11.5	0.863***	4.70	0.631	0.857***	1.52	−1.97
I/N	141	0.861***	2.12	−10.1	0.878***	4.53	0.652	0.871***	1.46	−1.83
R/C	56	0.827***	2.51	−15.2	0.846***	5.16	0.571	0.844***	1.70	−2.35
(F-value)			(3.1)	(4.5*)		(2.1)	(3.8)		(2.9)	(3.5)
Calf circumference (cm) on stature (cm)										
All	197	0.867***	2.64	−10.0	0.879***	4.14	0.847	0.875***	1.34	−1.45
I/N	141	0.872***	2.52	−8.4	0.881***	3.99	0.869	0.877***	1.28	−1.33
R/C	56	0.865***	2.93	−14.2	0.877***	4.53	0.791	0.876***	1.49	−1.77
(F-value)			(2.9)	(0.1)		(2.3)	(0.0)		(3.0)	(0.0)
Triceps skinfold thickness (mm) on stature (cm)										
All	185	0.585***	1.69	−15.4	0.622***	7.72	−0.191	0.615***	2.50	−4.47
I/N	129	0.576***	1.54	−13.3	0.610***	7.19	−0.111	0.604***	2.32	−4.08
R/C	56	0.609***	2.02	−20.2	0.644***	8.81	−0.353	0.636***	2.87	−5.27
(F-value)			(1.7)	(0.0)		(1.1)	(0.2)		(1.1)	(0.2)
Subscapular skinfold thickness (mm) on stature (cm)										
All	185	0.649***	4.09	−46.4	0.727***	13.87	−0.960	0.719***	4.49	−8.65
I/N	129	0.659***	3.80	−42.6	0.726***	13.32	−0.885	0.717***	4.29	−8.23
R/C	56	0.656***	4.82	−56.1	0.736***	15.20	−1.139	0.730***	4.98	−9.69
(F-value)			(1.8)	(1.9)		(0.8)	(0.6)		(1.0)	(0.5)

[a] Inland/northern group.

[b] Riverine/coastal group.

[c] Equality of slope and intercept by ANACOVA.

* p < 0.05; ** p < 0.01; *** p < 0.001.

Simple allometric analysis of human growth was applied to relationships between weight (or some other anthropometric measure) and stature (Sato, 1947; Shimizu and Inoue, 1956; Morishita, 1965, 1969; Kimura, 1970) and size relationships among somatic/cephalic characters (Shepherd et al., 1949; Sagami, 1967; Hoshi, 1978). Thompson (1942), on the other hand, found that a simple linear relationship (without transformation) was often adequate to describe the data, and since there is no a priori reason to prefer the allometric form and if the data show no curvilinear trend, then there is no justification for departing from a hypothesis of direct proportionality. As Tanner (1951) points out, there is no justification in the human species for a systematic transformation of the measurements into their logarithms. According to the result of the present analysis, however, the log weight/stature seems to yield clearer information than the simple linear (no transformation) and log-log (simple allometry) ones, which accords with the results in the Tutsi and Hutu of Rwanda (Hiernaux, 1964) and in the Isabela of the Philippines (Ashizawa et al., 1988).

Comparison of Relative Growth between the Sub-groups
Since the best fits to the trend lines were obtained in the semi-logarithmic plots (log weight/stature), comparisons between groups were performed with the slope (α) and intercept ($\log b$) of the regression line expressed as

$$\log y = \log b = \alpha x,$$

which is the logarithmic form of

$$y = b10^{\alpha x}.$$

Two regression lines, for inland/northern and riverine/coastal groups of the Gidra, are drawn for each measurement. According to the result of analysis of covariance (ANACOVA) (Tables 2 and 3), hypotheses of slope (α) equality between the two groups were not rejected in all the measurements for both sexes, excluding boys' triceps skinfold. Intercept ($\log b$) of weight was significantly different in boys and girls ($p < 0.001$). Since b represents the value of y at $x = 0$ and affects the initial size, it may be regarded as the initial growth index. Thus, the intra-population variation of weight to stature may occur before the age of subjects analyzed, namely in or before their infancy, and the variation continues through childhood and adolescence.

For the other measurements, only upper arm circumference in boys was significantly different ($p < 0.05$). Regression lines for boys display a tendency of superiority of the riverine/coastal group to the inland/ northern group except for calf circumference suggests that the size of calf circumference corresponds to different behavioral patterns; the people of the latter group travel only on foot, while the former travel either by canoe or on foot. Girls' figures are complicated to interpret since two regression lines in many measurements intersect. Most measurements against stature, however, are greater at the mature level in the riverine/coastal group than the inland/northern for girls as well as for boys, coinciding with the result on adult body physique (Chapter 11 in this volume).

CAUSAL FACTORS OF GROWTH VARIATION

Child growth was compared in the two groups of the Gidra population (i.e. inland/northern and riverine/coastal), and there were great differences in the growth of size (both stature and weight). According to the estimated growth curves, both sexes in the riverine/coastal group were taller and heavier than those in the inland/northern group. The constantly greater size of the former from the age of about three to 20 years provides evidence of earlier maturation in the former group. The corresponding result was obtained by the annual increment curves estimated by cubic spline functions. In terms of body shape (chest, upper arm, and calf circumferences), the two groups were not significantly different for all three circumferences of both sexes, but did differ for boys' upper arm circumference in relative growth to stature by means of a log value/stature case. From a consideration of group differences in stature, however, riverine/coastal children may have a greater magnitude in circumference for a given age (absolute growth).

Many factors that affect the rate of development are known; some are hereditary in origin, and others originate in the environment (Tanner, 1977: 339). Although genetic factors are clearly of importance and it is very difficult to specify quantitatively the relative importance of heredity, stature, for example, showing a strong genetic component, can be influenced in terms of growth rate and magnitude by environmental conditions. There are two basic approaches to investigations in which genetic variability is low and environmental differences great: on the one hand, migration studies (e.g. for Japanese, Greulich, 1957; for Africans, Hiernaux, 1963), and on the other, the present type of study of intra-population variation. The Gidra keep their cultural and social unity and no or few barriers to mating divide them, and these conditions minimize the possibility of a difference between the gene pool of the two sub-groups and highlight the environmental effects. The relevant ecological factors include diet, nutrition, climate, physiography, disease, and, perhaps, physical activity. Of these, nutrition must be first emphasized, as it may influence the development of bony tissue as well as soft tissue and operate to produce variation in developing and/or mature individuals.

The reports of the IBP (International Biological Programme) project on the Lufa and Karkar populations (in Papua New Guinea) give quantitative data on energy and nutrient intake in relation to child growth, in which nutrition has been regarded as the most important factor on growth patterns (Malcolm, 1969a, 1970a; Wark and Malcolm, 1969; Heath and Carter, 1971). The Gidra, whose intakes of energy and protein per adult male per day range from 2980 to 3553 kcal and from 54.3 to 73.3 g in the four villages (Chapter 9 in this volume), are taller through the growth period (Fig. 1) and as adults (Chapter 11 in this volume) than either the Lufa highlanders or the Karkar islanders, whose energy and protein intakes were 2520 kcal and 47.1 g, and 1940 kcal and 36.9 g, respectively (Norgan et al., 1974); the nutritional status of these two populations, and their ability to thrive, were better than could be expected from their low dietary intake (Ferro-Luzzi et al., 1975, 1978). The results of the comparison between the Gidra sub-groups on child growth (the riverine/coastal group was greater in size than the inland/ northern group) and on dietary intake (the former was lower in energy intake but higher in protein) reveal the importance of protein intake in the Gidra's relatively rapid growth and greater size. The difference in total protein intake between the two sub-groups is not due to the animal protein intake, though the proportion of aquatic animals to land animals differs with the ecological difference, but is due to the consumption of plant protein, especially that of purchased foods such as wheat flour and rice. Thus the growth pattern of the Gidra may reflect the change in food habits associated with newly introduced foods during the slow process of modernization.

Physical activity must be taken into account, especially in respect to energy expenditure (Chapter 6 in this volume). Inland/northern children have higher energy expenditure, since they travel on foot while the riverine/coastal children can use canoes. This effect of transit facilities on physical activities may affect not only growth rate but the shape of limbs. Indeed, a significant difference between the sub-groups is found in the relative growth of upper arm circumference for boys (Table 2). In calf circumference, inland/northern subjects were similar to (for boys) or greater than (for girls) the riverine/coastal children.

A striking tendency toward acceleration in growth, known as the "secular trend," has been reported in industrialized countries (Tanner, 1962, 1968; Kimura, 1967; Ljung et al., 1974; Greulich, 1976; Meredith, 1976; Frisancho et al., 1977; van Wieringen, 1978). This trend may come to an end in the developed countries, where nutrition has become near optimal for growth (Damon, 1968; Kimura, 1977; van Wieringen, 1978; Tanner, 1978; Suzuki, 1981). In the Gidra a secular trend appeared in the growth pattern, though the slight stature difference between adults and elders was judged to be a true loss of height rather than a secular trend (Chapter 11 in this volume). That is because the traditional way of life of the Gidra is changing little by little; for example, the food taboo system related to clan organization has relaxed and is maintained only by the aged. Intra-population variation in the secular trend may be occurring in association with the degree of change in the way of life, since the riverine/coastal children, who are more affected by modernization through consumption of purchased food and the enlargement of horticulture (Chapter 9 in this volume), are larger in size and mature earlier than the inland/northern children. Changes in food consumption patterns will affect physical growth of the younger generation in the future.

(*Toshio Kawabe*)

Chapter 14

Blood Pressure and Obesity

In human populations, a high salt (NaCl) intake is associated with a high prevalence of hypertension (Dahl, 1972; Tobian, 1979), but the critical level of Na intake for incidence of hypertension in individuals and in populations is debatable (Freis, 1976). In populations subjected to modern life with NaCl intakes about 15–20 g/day, an age-related increment of blood pressures is commonly observed (Hamilton *et al.*, 1954; Miall and Chinn, 1973), while in traditional populations with NaCl intakes less than 5g/day, no rise in blood pressure was observed with aging (Prior *et al.*, 1968; Boyce *et al.*, 1978). Thus, the level of 5 g NaCl/day may be critical in the genesis of hypertension with aging. However, there are many factors other than NaCl intake related to the incidence of hypertension. For example, obesity is an important factor (Tyroler *et al.*, 1975), and it is associated with high NaCl consumption in modernized populations (Miall *et al.*, 1968; Kannel *et al.*, 1967). This together with genetic differences in individual sensitivity to NaCl complicated the problem (Altshul and Grommet, 1980).

Therefore, it is meaningful to elucidate the effect of NaCl consumption on blood pressure among a Gidra population consisting of several subgroups with different level of NaCl intake but otherwise with similar living and genetic conditions. As reported previously (Chapter 9 in this volume), the consumption of foods hitherto not available in this area has spread in recent years as a cash economy was adopted, though not to a full extent. This makes a difference in salt consumption among villages. Traditional salt of the Gidra was sodium-rich tree ash, and consumption was rather limited to a small extent (Ohtsuka *et al.*, 1986). Nowadays, the Gidra can buy table salt imported from foreign countries. It was possible therefore to select villages with different levels of salt consumption.

THE SUBJECTS

As in other chapters in this volume, four villages were selected: Rual in the north (inland), Wonie inland, Ume on the bank of the Binaturi River, and Dorogori on the coast. Among the inhabitants in the four villages children and adolescents were excluded in this study. As for convenience of transportation to the town, the coastal village, Dorogori, occupies the best position in terms of access to Daru town (only 1–2-hr trip by sailing canoe); the second best is the riverine village, Ume (about six hours by canoe equipped with engine), and then the central inland village, Wonie (1–2 days by walk and motorized canoe), and the farthest inland village, Rual (2 days by walk and motorized canoe). Their accessibility to Daru is related to the frequency of purchasing imported foods (e.g. rice, flour, and tinned fish) which contain more salt than local foods, and table salt. From the results of our household food consumption survey in 1981 (Chapter 9 in this volume) and the chemical analysis of the food in the area (Chapters 7 and 8 in this volume), and of drinking water (Ohtsuka et al., 1985), the average daily intake of Na was estimated for each village; the highest was 1.8 g/person in Dorogori, the next highest, 1.5 g/person in Ume, then 1.1 g/person in Rual, and the lowest 0.8 g/person in Wonie (Chapter 10 in this volume). In the 1971–72 survey in Wonie (Ohtsuka and Suzuki, 1978), Na consumption was estimated to be around 0.2 g/day, which almost exclusively came from foods. Therefore, the increased consumption of Na was quite conspicuous.

METHODS

Estimation of Salt Consumption

The individual's Na consumption was estimated using the data urinary excretion of Na. Urinary Na, K, and creatinine (Cr) were measured for the first morning (overnight) urine which was collected in a plastic cup from 231 villagers (105 males and 126 females) in December 1981. A sheet of filter paper (4.5 cm × 2 cm) was dipped into the sample of urine, and after drying about one hour in a room, it was packed with aluminum foil to be brought to Japan. By this sampling technique, urinary Na, K, and Cr levels remained constant in the samples absorbed on the filter paper after storage for several months (Takemori, 1980). After, using Jaffe's reaction, extracting urine from the filter paper with distilled water (for Na and K) or diluted HCl solution (for Cr), Na and K were measured by flame photometry, and Cr by colorimetry.

In this study, daily salt consumption was assessed by daily urinary Na excretion, which was estimated from the level of Na per g creatinine (Cr) in the morning urine and daily urinary excretion of Cr based on individual's fat-free mass from anthropometric data (Forbes and Bruining, 1976). To validate the procedure, 24-hr urinary excretion of Na was compared with the above-mentioned estimation for samples obtained from 30 adults in Dorogori village. The amount of 24-hr urinary Na excretion was significantly correlated with the amount of Na in morning urine, estimated from Na/Cr ratio and the amount of Cr ($r = 0.53$, $p < 0.01$), and the means of the two values were not statistically different ($p > 0.05$).

Body Fat

One of our research team members (T.K.) conducted anthropometry in July and August 1981; details of the anthropometric study are reported in Chapter 11 in this volume. From skinfold values measured at triceps and subscapular sites using a Holtian skinfold caliper, the body density was calculated by the equation of Nagamine and Suzuki (1964), and then the body fat percentage was calculated by the equation of Brožek *et al.* (1963). Fat-free mass was thus calculated from body weight and fat percentage.

Blood Pressure

Blood pressures were measured in August 1981 by one of our research team members (T.S.) with the subject at sitting position. With a mercury sphygmomanometer, the appearance and the disappearance of the sounds were determined as systolic blood pressure (SBP) and diastolic blood pressure (DBP), respectively. Two to three trials were made and the lowest reading to the nearest 5 mmHg was recorded. Examined were 205 villagers (97 males and 108 females). Because the urine sampling was done in December 1981, the blood pressures were measured again in the same month for Dorogori villagers to examine the validity of using the August data in relation to the urinalysis data; the two blood pressure data sets were significantly correlated ($p < 0.05$).

RESULTS

Salt Consumption

Table 1 shows estimated daily urinary excretion of Na and K by village and by sex/age group. The average urinary Na excretion was slightly higher, but not statistically significant, in Ume and Dorogori villagers than in Rual and Wonie villagers. The individual Na value was very variable within each village. Quite surprisingly, very low Na values (e.g. 30 mg/day) were recognized in some elder cases in Rual village, although the reason is not clear at present.

The mean of urinary Na excretion almost coincided with the daily Na

Table 1. Daily Urinary Excretions of Na and K (mean±SD in mg/day)[a]

Sex/age group		Rual	Wonie	Ume	Dorogori	Total
Na	Male adults	705±770	690±730	955±660	1320±755	920±740
		(15)	(19)	(30)	(16)	(80)
	elders	130±145	145±145	2080±1495	630±380	980±1220
		(2)	(3)	(6)	(7)	(18)
	Female adults	485±565	575±560	1055±835	920±730	810±740
		(16)	(20)	(33)	(19)	(88)
	elders	30	585±780	920±855	765±585	785±740
		(1)	(3)	(11)	(7)	(22)
K	Male adults	2570±1440	2840±2150	2775±1490	3085±1885	2815±1715
		(15)	(19)	(30)	(16)	(80)
	elders	1490±820	1770±780	4175±2345	3335±1020	3150±1775
		(2)	(3)	(6)	(7)	(18)
	Female adults	2175±910	2895±1930	2420±1345	3015±1645	2610±1510
		(16)	(20)	(33)	(19)	(88)
	elders	910	3285±1620	2325±970	4035±1970	2935±1620
		(1)	(3)	(11)	(7)	(22)
Na/K	Male adults	0.55±0.49	0.69±1.22	0.74±0.53	0.97±0.61	0.74±0.74
		(17)	(19)	(32)	(18)	(86)
	elders	0.12±0.10	0.12±0.08	1.02±0.92	0.35±0.23	0.54±0.67
		(2)	(3)	(7)	(7)	(19)
	Female adults	0.38±0.48	0.52±0.80	1.03±1.62	0.66±0.76	0.72±1.16
		(19)	(20)	(37)	(21)	(97)
	elders	0.30±0.34	0.24±0.35	0.74±0.69	0.38±0.24	0.48±0.51
		(6)	(4)	(11)	(8)	(29)

[a] The number of cases is shown in parentheses: the number of cases for the Na/K ratio is, in most instances, greater than that for either Na or K, since the daily excretion of Na and K could not be obtained for subjects without anthropometric measurement (lack of fat-free mass values). Means of urinary Na, K, and Na/K did not vary with village for each sex/age group.

intake of Wonie villagers as shown in Table 1; urinary Na outputs in male and female adults were estimated as 690 and 575 mg/day, respectively, and Na intake per adult was 730 mg/day. In male adults of the other three villages, however, the average urinary Na excretion was lower than daily Na intake.

As for K, the inter-village difference of mean urinary excretion was small compared with the case of Na. Therefore, urinary Na/K showed comparable patterns of inter-village differences to urinary Na output.

Body Fat

The means and SDs of stature, weight, sum of two skinfold values, body mass index [weight (kg)/stature (m) squared, BMI], body fat percentage (% fat), and fat-free mass (FFM) are similar to the values reported in Chapter 11 in this volume; there are, however, minor differences because

Table 2. Blood Pressures by Sex and Age Group (mean ± SD)[a]

Sex/age group		Rual	Wonie	Ume	Dorogori	Total
SBP[b]	Male adults	122±13	117±11[A]	129±11[AB]	119±8[B]	123±12
		(15)	(19)	(29)	(16)	(79)
	elders	128±18	123±30	130±14	128±22	128±19
		(2)	(3)	(6)	(7)	(18)
	Female adults	122±10	118±12	128±17	124±18	124±15⌐
		(16)	(20)	(31)	(19)	(86) ⌐*
	elders	125	125±23	145±14[A]	129±9[A]	136±16⌐
		(1)	(3)	(11)	(7)	(22)
DBP[c]	Male adults	70±9	68±8	74±10[A]	65±8[A]	70±9
		(15)	(19)	(29)	(16)	(79)
	elders	73±4	70±13	70±8	67±9	69±8
		(2)	(3)	(6)	(7)	(18)
	Female adults	73±8	68±9	74±10[A]	65±14[A]	71±11
		(16)	(20)	(31)	(19)	(86)
	elders	70	68±18	78±12	60±15	70±15
		(1)	(3)	(11)	(7)	(22)
PP[d]	Male adults	52±9	49±10	55±13	54±9	53±11
		(15)	(19)	(29)	(16)	(79)
	elders	55±21	53±19	60±13	61±19	59±16
		(2)	(3)	(6)	(7)	(18)
	Female adults	48±8	50±10	54±13	59±15	53±12⌐
		(16)	(20)	(31)	(19)	(86) ⌐*
	elders	55	57±8	67±13	69±13	66±12⌐
		(1)	(3)	(11)	(7)	(22)

[a] The number of cases is shown in parentheses. Means with same character differ with each other ($p < 0.05$) by Scheffe's test for multiple comparison. [b] Systolic blood pressure. [c] Diastolic blood pressure. [d] Pulse pressure.
*$p < 0.05$.

the values for some villagers who did not participate in the blood pressure measurement or urine sampling were excluded in this analysis. Among males, elders showed significantly lower values in weight, BMI, and FFM than adults (each, $p < 0.05$). In male adults, body weight, skinfolds, BMI, % fat, and FFM were higher in Dorogori and Ume villagers than in Rual and Wonie villagers.

Female adults had inter-village differences of physique similar to those of male adults, showing the highest body weight, BMI, and FFM in Dorogori people.

Blood Pressure

Table 2 indicates blood pressures by village and by sex/age group. In females, SBP and pulse pressure (PP) were higher in elders than in adults. Noticeable was that blood pressures were generally high in Ume villagers

Table 3. Correlation Coefficients of the Blood Pressure with Their Physique and Urinary Na and K by Sex and Age Group

	Males					
	Adults (N=79)			Elders (N=18)		
Variable	SBP[b]	DBP[c]	PP[d]	SBP	DBP	PP
Height					0.46*	
Weight						
Skinfolds[e]						
Na[f]						
K[g]			−0.27**			
Na/K						

	Females					
	Adults (N=86)			Elders (N=22)		
Variable	SBP	DBP	PP	SBP	DBP	PP
Height						
Weight			0.26*			
Skinfolds	0.33***		0.36***	0.33 (p=0.06)	0.27 (p=0.11)	
Na						
K						
Na/K						

[a] The correlation coefficient is shown only in cases of correlation at significant or marginal level. [b] Systolic blood pressure. [c] Diastolic blood pressure. [d] Pluse pressure. [e] Sum of two skinfolds (triceps and subscapular). [f] Daily urinary Na excretion. [g] Daily urinary K excretion.
* $p < 0.05$, ** $p < 0.01$, *** $p < 0.001$.

than in Dorogori and Wonie villagers. Also, DBP of female adults and SBP of female elders were higher in Ume people than in Dorogori people.

Table 3 compiles simple correlation matrices between each of the blood pressure measurements and each of three physique and three urinary measurements in each sex/age group. Three measurements of body physique were selected from principal components analysis; the loading of the first factor was high in the sum of two skinfolds and FFM, and that of the second factor was high in stature, weight BMI, and %fat. In male adults, PP correlated inversely with urinary excretion of K, and in male elders, DBP correlated positively with stature. In female adults, positive correlations were observed between SBP and skinfold thickness, and between PP and both body weight and skinfold thickness. Female elders, whose SBP was the highest among the Gidra, showed marginal positive correlations of SBP and DBP with skinfold thickness.

DISCUSSION

In Wonie village, the Na intake per adult male from foods increased from 200 mg/day in 1971–72 (Ohtsuka and Suzuki, 1978) to 360 mg/day in 1981. The increase of Na intake was caused by a change in consumed foods associated with the introduction of Western salty foods and addition of table salt. The distance from each of the four villages to Daru town, or the accessibility to foreign foods, corresponded to the level of Na intake. Noticeable is the fact that the average NaCl intake was still lower than 5 g/day, which is claimed to be critical for induction of hypertension with aging (Dahl, 1972; Freis, 1976), even in the most modernized Gidra village.

In order to estimate individual daily salt consumption, we adopted the following assumptions: the Na intake was roughly identical to urinary Na excretion; Na/Cr in the morning urine was representative of Na/Cr in 24-hr urine; daily urinary Cr excretion could be predicted from FFM values. Obviously, this procedure of estimation includes several error sources. Actually, the correlation coefficient between the measured and estimated daily Na excretion was 0.53, as mentioned. This has certainly created a weakness in the present study, and moreover, the large variation of urinary Na excretion has made the inter-village comparison insignificant, although the trend observed (higher Na intake in riverine/coastal villagers than in northern/inland villagers) in our food consumption survey was recognized.

In the Gidra, significant inter-village differences of body physique except for stature were observed (Chapter 11 in this volume). Two village groups—riverine/coastal group and northern/inland group—were distinguished in male and female adults. The former group was fatter than the latter. The riverine/coastal group took more foreign foods with high fat content than the northern/inland group, even though the average energy intake was rather greater in the latter (Chapter 9 in this volume). However, daily energy expenditure seemed to be lower in the riverine/coastal group than in the norther/inland group since the energy cost of the transportation of the former (by canoe) was less than that of the latter (by walking). From the data indicating body fatness (skinfold values, %fat, and BMI), relatively excess energy intake is most likely to be in the coastal/riverine people.

Sex/age comparison of the Gidra blood pressures demonstrated that SBP was higher in female elders than in both male elders and female adults; the similar sex/age variation was reported from several traditional societies (Prior et al., 1968; Truswell et al., 1972; Sinnett and Whyte, 1973; Page et al., 1974). For this explanation, postmenopausal change of hormonal activities (Page *et al.*, 1974) and selective mortality of hypertensive male elders (Sinnett and Whyte, 1973) have been proposed, although these factors could not be examined for our Gidra data. One of the conceivable reasons for the high blood pressure of female elders was that they had slightly thicker skinfolds than female adults despite elders' significantly lighter FFM (cf. Chapter 11 in this volume).

The inter-village difference of blood pressures was characterized by higher SBP and DBP in riverine (Ume) villagers (Table 2). Examining several factors associated with blood pressure by the simple correlation analysis (Table 3), we found the skinfold as a contributing factor. Puzzlingly, however, the largest skinfold values were observed in the coastal (Dorogori), but not riverine (Ume) villagers. Thus, some additional factors have to be sought for the higher blood pressures in riverine villagers.

In summary, in accordance with modernization, the Na intake of the Gidra has increased. Elevation of blood pressures with aging was found only in females. This was associated not with the increased Na intake but with gains in fat, although the fat level did not reach a level judged as obese (cf. Chapters 3 and 11 in this volume). One may suppose that two following factors may be related to the variation of the Gidra blood pressures.

First, some researchers found that lower blood pressure was associated with poor health, higher splenomegaly and hepatomegaly (Burns-Cox, 1970; Page *et al.*, 1974). However, most of the Gidra were healthy and they seemed to be free from hepatic and renal diseases, because detectable amounts of urobilinogen, protein, and glucose were found in 6.0, 4.6, and 0.9%, respectively, of the adults and elders (unpublished data), and palpable splenomegaly was not found.

Second, other dietary factors including K and Ca intakes may inversely related to blood pressure (Meneely and Battarbee, 1976; Calabrese *et al.*, 1980; Khaw and Barrett, 1984). However, we found no inter-village differences in urinary K excretion, although DBP of male adults showed a weak inverse association with urinary K excretion. Drinking water in the coastal village contained the highest Ca among the villages (Ohtsuka *et al.*, 1985; also see Chapter 10 in this volume), and this might contribute to some extent to the lower blood pressure in coastal villagers than in riverine villagers. Further studies are needed to clarify the role of K, Ca, or other dietary factors on the blood pressure regulation in the Gidra.

(*Tsukasa Inaoka*)

Chapter 15

Urinalysis and Protein Nutriture

Papua New Guineans have been characterized by marginal nutritional conditions. Hipsley and Clements (1950) reported 1600 kcal of energy intake and 22 g of protein intake per day per person among the sweet potato eaters in the highland, and the studies followed (e.g. Oomen, 1961; Covee et al., 1962; Luyken et al., 1964; Fujita et al., 1986) also stressed their low protein intake, although all of the subjects were highlanders and perhaps were selected from the less-modernized regions. In contrast, the Gidra people showed much higher energy and protein intakes. The major reasons for their high energy intake were discussed in relation to energy expenditure (Chapter 6 in this volume). Their high protein intake was attributed to the rich fauna in the thinly populated Gidraland and the recent increase of consumption of purchased foods, judged from the subsistence and food consumption patterns (Chapter 9 in this volume).

Physiologically, the ingested protein passes through the body nitrogen pool and is excreted mainly in the urine, with small amounts of loss in the feces and through the skin (Simmons, 1972). Nitrogen compounds in the urine exist mainly in the form of urea nitrogen (UN), and its amount tends to fluctuate in accordance with protein intake (Yamori et al., 1982 b). In contrast, urinary creatinine (Cr) reflects endogenous nitrogen loss, and its level is affected less by dietary intake than by body composition (Heymsfield et al., 1983). This chapter aims first to elucidate protein nutriture of the Gidra people from UN and Cr values of urine samples collected in 1986. In this connection, Bohdal et al. (1968) pointed out that inorganic sulfate sulfur (SO_4) or SO_4/Cr would be a good indicator for protein intake and particularly animal protein intake, instead of UN or UN/Cr, although the feasibility of the analysis of SO_4 is still being examined, and the SO_4 (or S) concentrations of our urine samples will be reported later. Second, sodium (Na) and potassium (K) concentraions of the same urine samples are considered in relation to the urinary UN and Cr values on the one hand, and on the other are compared to the

Na and K concentrations of the samples collected in 1981 (Chapter 14 in this volume).

METHODS

During the patrol survey period from September to December 1986, I collected urine samples in our intensively studied villages (Dorogori, Ume, Wonie, and Rual); in the same duration the people's energy expenditure by means of heart rate method was also examined (Chapter 6 in this volume). The number of subjects in the four Gidra villages was 470, which exceeded 90% of the total inhabitants. For comparative purposes, urine samples were also collected from 35 Gidra people, of various ages and of both sexes, in Daru town, 18 (ten male and eight female) Gidra highschool students in Daru, and 16 unmarried males in a coastal Kiwai-speaking village, eastern adjacent to Dorogori. As in other analyses of our Gidra studies, these subjects were grouped according to their age-grade system, but recategorized into unmarried and married groups because of the small number of aged subjects; as a result, the subjects numbered 159 for unmarried males, 95 for married males, 101 for unmarried females, and 115 for married females.

The first morning (overnight) urines were sampled for the advantage of the people's participation. The methods of collecting urine, which was adsorbed in filter paper, and of transporting the samples were the same as those in 1981 (see Chapter 14). It has already been demonstrated that when this technique was used, urinary levels of not only Na and K but also Cr and UN remained constant after storage in a freezer for several months. After extracting urine from the filter paper with diluted HCl solution, Na and K concentrations were measured by flame photometry, that of Cr by Jaffe's procedure, and that of UN by the urease method.

Whether urinary constituents from 24-hr urine samples are represented by partial urine samples has been debated. For instance, Simmons (1972), examining UN concentrations of his own and Arroyave and Lee's (1966) samples, pointed out that the best fit to 24-hr samples was found in the partial urines collected in the morning after the first voiding. However, Yamori et al. (1982a) disclosed that in any constitutents, i.e. UN, Cr, UN/Cr, Na, K, and Na/K, any of the four partial specimens (first morning, morning, afternoon, and evening) were significantly correlated with the 24-hr specimens, and that UN and UN/Cr in the first morning urines showed particularly good correlations with those in 24-hr urine, with the correlation coefficients of over 0.75. My own comparison between the urinary constituents in the morning samples and those in the 24-hr samples for 30 Gidra villagers in 1981 proved significant correlations in any constituents (cf. Chapter 14 in this volume).

Measurements other than urinary constituents, which were analyzed in this study, included systolic and diastolic blood pressures (SBP and

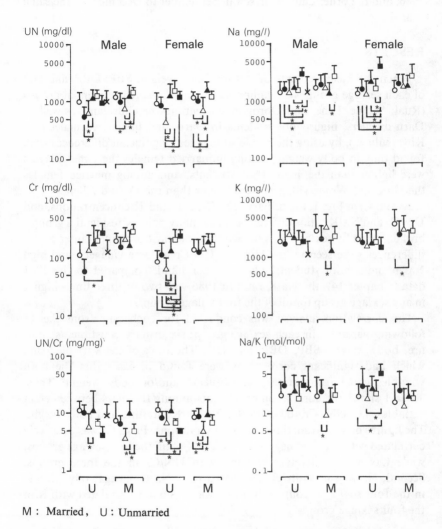

M : Married, U : Unmarried

Fig. 1. Inter-group differences of urinary constituents. Significantly different pairs of study-site by Sheffe's range test are also shown (*, p < 0.05). ○: Rual village, ●: Wonie village, △: Ume village, ▲: Dorogori village, □: Daru dwellers, ■: high-school students, × : Kiwai village.

DBP) and heart rate (HR) at sitting position, all of which were measured in the urine collecting period, using an automatic digital sphygmomanometer, and stature and body weight data; the anthropometric data were obtained by one of our research team members (T.K.) in July and August 1986, and the other data by myself in September to December of the same year.

RESULTS

When the measurements of stature, body weight, SBP, DBP, and HR of each sex/age group were compared among the seven residential groups (Rual, Wonie, Ume, and Dorogori villagers in the Gidraland, as well as Daru dwellers, highschool students in Daru, and unmarried males in a Kiwai village), by using the Sheffe's range test, significant differences were found only in body weight; among unmarried females the Ume villagers were lighter than the highschool students, and among married females the Ume and Wonie villagers were lighter than the Daru dwellers.

As shown in Fig. 1, the urinary UN, Cr, Na, and K concentrations, and UN/Cr and Na/K ratios differed between many pairs of residential groups, broken down by sex/age. It is judged that the most prominent inter-group differences were seen in low UN and UN/Cr in Ume villagers, and high Na in highschool students. On the other hand, compared to the 1981 data (Chapter 14), the Na/K ratio in 1986 was two to three times higher in any sex/age group for all of the four village groups.

The correlation analysis was conducted for each pair from the 11 following variables in each sex/age group: six urinary constituents, stature, body weight, SBP, DBP, and HR. The bulk of the variable pairs which had significant correlations were found in cases that the both variables came from urinary constituents and/or body weight. Thus, Table 1 shows the correlation coefficients among the pairs between eight variables, i.e. UN, Cr, UN/Cr, Na, K, Na/K, stature, and body weight. The figures demonstrate three major observations. First, Cr was positively correlated with UN and negatively with UN/Cr in the four sex/age groups, while UN was positively correlated with UN/Cr in the three groups. Second, Na/K was positively correlated with Na and negatively with K in the four sex/ age groups. Third, Cr was positively correlated with K in the four sex/age groups.

PROTEIN NUTRITURE

Experimental studies (Yamori *et al.*, 1982a; Fujita *et al.*, 1986) have revealed that the excretion levels of nitrogen compounds in urine reflect

Table 1. Correlation Matrices by Sex and Age Group

Unmarried Male (N=159)

	UN	Cr	UN/Cr	Na	K	Na/K	Stature
Cr	.56***						
UN/Cr		−.41***					
Na							
K		.40***	−.25**	.18*			
Na/K		−.23**	.20*	.32***	−.48***		
Stature	.22**	.47***	−.35***				
Weight	.27**	.53***	−.34***			.20*	.94***

Married Male (N=95)

	UN	Cr	UN/Cr	Na	K	Na/K	Stature
Cr	.43***						
UN/Cr	.41***	−.46***					
Na		.22*					
K		.50***	−.36***	.33**			
Na/K		−.36***	.27**	.27**	−.54***		
Stature							
Weight							.65***

Unmarried Female (N=101)

	UN	Cr	UN/Cr	Na	K	Na/K	Stature
Cr	.79***						
UN/Cr	.22*	−.30**					
Na	.35***	.52***					
K	.22*	.31**					
Na/K				.44***	−.51***		
Stature		.28*					
Weight	.39**	.44***		.27*		.24*	.91***

Married Female (N=115)

	UN	Cr	UN/Cr	Na	K	Na/K	Stature
Cr	.53***						
UN/Cr	.51***	−.29**					
Na							
K		.28**	−.23*				
Na/K		−.29**		.30**	−.59***		
Stature							
Weight	.24*			.20*			.55***

* $p<0.05$, ** $p<0.01$, *** $p<0.001$.

Figures are given only for the significant correlation coefficients ($p<0.05$).

dietary protein intake levels, not of the preceding one or two days but of the preceding several days. In the present study, the foods consumed in the day before the urine sampling were investigated by the interview method. When the UN and UN/Cr values were compared between the subjects who had taken protein-rich foods in the previous day and those who had not, either UN or UN/Cr value was slightly but not significantly higher in the former than in the latter. Thus, the UN and UN/Cr values of the present subjects may have reflected the longer-duration protein intake levels.

To evaluate protein nutriture from urinalysis, UN/Cr ratio was judged to be the most suitable measure (Simmons, 1972). The UN/Cr of the present subjects ranged approximately from 5 to 10 $\mu g/\mu g$. Analyzing the relationship between UN/Cr ratio and daily protein intake per unit body weight among three experimental groups whose protein intakes differed from 0.5 to 2.5 g/kg body weight (Arroyave and Lee, 1966) and two Mountain Ok-speaking village groups in highland Papua New Guinea, whose per-day protein intake was as low as 0.4–0.5 g/kg body weight (Ohtsuka et al., unpublished data), the two variables demonstrated a linear relationship; 1 g of per-day per-body weight protein intake corresponded to about 7 $\mu g/\mu g$ of UN/Cr. When the present data were applied to this relation, the protein intake per 1 kg body weight was estimated at 0.7–1.5 g. Though we did not conduct a food consumption and nutrient intake survey in 1986, the protein intake of the Gidra villagers in 1981 (Chapter 9 in this volume) were about 1.0–1.3 g/kg body weight, falling in the above range. Thus, it can be concluded that the UN/Cr ratio in the present samples well reflected the Gidra people's protein intake level and that this level is much higher, compared to highlanders in Papua New Guinea.

In considering the low UN/Cr level of Ume villagers, we should note that UN concentration was correlated with body weight among three sex/age groups though not among married males (Table 1), and that Ume villagers were lighter than the other village groups, though the significant differences were found only in unmarried females in comparison with the highschool students and in married females in comparison with their Daru-dwelling counterparts. Since our anthropometric study among the Gidra in 1981 (Chapter 11 in this volume) showed that adult males' body weight was heavier in Ume than in the other villages, and body weight of adult females in Ume was comparable to that in Rual and Wonie, the ecological situation of Ume in the duration from 1981 to 1986 should be examined.

There were two problems which may have been involved in the decrease of body weight in Ume villagers. The first was the prevalence of taro blight caused by a virus, *Phytophthora colocasiae*, which has been widespread in Papua New Guinea and other Pacific regions (Purseglove, 1972; Bourke, 1982). In the Gidraland, this disease was not known in the past and has become severe only since 1985, and because Ume villagers had heavily depended on taro they had the most serious damage.

The second problem came from the political disputes over landowning rights with a neighboring linguistic group, and resulted in urging the Ume villagers to move their village to a new site; in fact, the village people could not reach a consensus on forming a single village but were split into two groups. Because of the short time period for preparing the village move, their horticultural gardens were still less productive in 1986. These two matters took place in accordance, more or less, with the modernization process. As discussed in Chapter 9 in this volume, modernization tended to raise the Gidra people's nutritional status, but it may also have drawbacks.

SODIUM AND POTASSIUM CONCENTRATIONS

One of the interesting results in the correlation analysis of the six urinary constituents was that K concentration was significantly correlated with Cr concentration in all sex/age groups. According to the previous studies, Cr was correlated with lean body mass (LBM) (Forbes and Bruining, 1976; Heymsfield et al., 1983), and LBM correlations with K as well as Cr were also reported (Muldowney et al., 1957). Thus, it is considered that both Cr and K of the Gidra subjects were related with LBM and consequently that both concentrations were interrelated, although our anthropometric study in 1986 did not include the measurements of skinfold thickness and thus failed to estimate their LBM.

When we compared the results in our 1981–82 study, Na/K ratio was two to three times higher in all sex/age groups in the four villages (Chapter 14 in this volume). The direct reason for this change came from the increase of Na concentration. Nonetheless, the daily Na intake level which was estimated from the urinary Na concentration was less than 5 g as NaCl, the level which was claimed to induce hypertension (Dahl, 1972; Freis, 1976). Thus, it can be concluded that the Gidra people increased their Na intake in the five-year period but that their Na intake has still remained at a lower level than that which may elevate blood pressure to hypertensive levels.

The high urinary Na concentration among the highschool students was prominent. The students resided in the school dormitories and ate most of their foods in the dormitory messroom. According to our interview survey about their food consumption pattern, their menu had very little variety; regularly served were rice, wheat flour, and a small amount of tinned fish, and their dishes contained a large amount of salt. According to our interview survey with the Daru-dwelling Gidra people, their food consumption pattern was not similar to that of the highschool students, but they ate many more foods grown in the villages, carried from their birth villages, or purchased in the local market in Daru. However, when the change in food consumption pattern as seen among the highschool students progresses in the course of future modernization process, their Na intake will reach the level triggering hypertension; in a broader sense, the incidence of cardiovascular disorders will rise like that in developed countries.

(Tsukasa Inaoka)

Chapter 16

Micronutrients in Hair

In evaluating nutritional status of trace elements, one effective approach is to measure or to estimate people's dietary intake of these elements. However, this approach has a methodological drawback, in particular, the lack of assessment on the bioavailability of elements, which markedly varies according to dietary food composition. Among various biological media for nutritional assessment on trace element status, hair is easy to obtain even from donors in traditional societies, and it can be stored for a long period.

Our first study of element concentrations of Gidra people's hair was conducted in 1971–72 for Wonie villagers, and disclosed that Na and Zn concentrations were markedly low compared to those in Japanese (Ohtsuka and Suzuki, 1978; Sasaki et al., 1981). After a ten-year interval, 505 hair samples were collected from four villages of the Gidra in 1981–1982, and concentrations of 12 trace elements (Na, Mg, Al, P, K, Ca, Mn, Fe, Cu, Zn, Sr, and Pb) were measured; hair Hg concentration was also analyzed and reported separately (Chapter 17 in this volume).

The scalp hair is very susceptible to exogenous contamination by atomospheric pollutants, water, sweat, and cosmetics (Hongo et al., 1988; Rivlin, 1983; Chittleborough, 1980; McKenzie, 1978; Hambidge, 1973); washing of the hair results in different degrees of elimination according to the elements. For instance, Cu, Zn, Hg, and Pb are least likely to be washed out, while Na, K, Mg, Ca, and Sr are very likely to be eliminated by washing (Suzuki, 1988; Suzuki et al., 1984b). In non-washed hair, elemental levels reflect both intake level and exogenous contaminations (Chittleborough, 1980). Comparing elemental compositions of non-washed hair with elemental intake (Chapter 10 in this volume) and elemental compositions of water (Chapter 10 in this volume; also see Ohtsuka et al., 1985), this chapter aims to consider the Gidra people's nutritional and health status.

METHODS

Five-hundred-five hair samples were collected in the period from August to November 1981 in four villages: Rual, Wonie, Ume, and Dorogori. Hair samples were taken from various parts of the head with stainless-steel scissors on the occasion of hair-cutting by the villagers themselves, and kept in polyethylene bags.

After being transported to the laboratory in Japan, approximately 0.15 g of non-washed hair samples were weighed and heated for 3 hr at 120°C with addition of 3 ml of nitric acid in a Teflon-lined, high-pressure decomposition vessel (Uniseal Decomposition Vessels, Ltd.). Then the digested samples were diluted to 10 ml with deionized distilled water.

Sodium, Mg, Al, P, K, Ca, Mn, Fe, Cu, Zn, Sr, and Pb concentrations in the diluted samples were measured by Inductively Coupled Plasma Atomic Emission Spectrometry (ICP-AES, Model 975 Plasma Atomcomp, Jarrelash). The accuracy of the determination was checked by measuring a reference human hair sample (National Institute for Environmental Studies of Japan, NIES No. 5).

For statistical analysis, element concentrations in hair samples were converted to logarithms. Difference in element concentrations between sexes and between age groups (two groups, i.e. adults or (ever-) married subjects and children or unmarried subjects) was tested by Student's t-test. Variation in element concentrations by villages was analyzed by one-way analysis of variance (ANOVA), and difference between the average values among the villages was tested by Duncan's multiple range test when a significant difference was noted by ANOVA. Principal com-

Table 1. Comparison of Hair Element Concentrations of the Gidra in the Range of the Average Value in Literature

Element	Elimination rate by acetone/water washing (%)	Mean level of the Gidra (μg/g) Level in non-washed hair	Estimated level after washing	Reported levels (acetone/water washing, μg/g)
Na	97.5	989	25	3.1–153
Mg	39.2	397	241	36.1–156
Al	50.9	336	165	5.9–32
P	17.5	159	131	105–151
K	94.0	701	42	1.5–33
Ca	44.4	1517	843	452–1840
Mn	23.3	38.4	29.5	0.4–7.3
Fe	45.2	515	282	14–90
Cu	6.8	11.8	11.0	9.3–20
Zn	0.8	121	120	124–248
Sr	37.1	11.7	7.4	*
Pb	0.1	0.62	0.62	1.3–13.2

* no data available.

ponents analysis with varimax rotation was applied to reveal the inter-relationship of element concentrations. All statistical analyses were conducted using the SPSS-X program package (SPSS-X™ User's Guide, 1988).

This study treats non-washed hair element concentrations, although there are very few such data in the literature. In order to compare the present results with those of previous studies, we conducted additional experiments on the elimination of elements from 57 Gidra hair samples by washing with acetone and water. Using the elimination rates, the hair element concentrations after washing were compared with the range of the average values in the literature (Imahori *et al.*, 1979; Takeuchi *et al.*, 1982; Fukushima *et al.*, 1982; Bhat *et al.*, 1982; Takagi *et al.*, 1986; Moon *et al.*, 1986) (Table 1). Of the 12 elements analyzed for hair samples, Pb showed concentration below the detection limit (5 μg/g) in 470 subjects (93%). Thus, in calculation of the mean Pb value, one-tenth of the detection limit, i.e. 0.5 μg/g, was used for such cases. Concentrations of Mg, Al, K, Mn, and Fe in the Gidra hair were higher than the reported levels, while those of Zn and Pb were slightly lower.

INTER-VILLAGE COMPARISON

Comparing the mean levels of element concentrations among groups broken down by sex and age group, differences were not so marked but were statistically significant in several elements. Concentrations of Al, K, and Mn were higher in females and that of Zn was higher in males.

Table 2. Geometric Mean and Its Standard Deviation of Hair Element Concentrations in the Four Villages (μg/g)

Element	Village				Difference by ANOVA
	Rual	Wonie	Ume	Dorogori	
Na	719(2.1)[B]	634(1.9)[B]	679(2.1)[B]	3750(2.0)[A]	p<0.001
Mg	168(1.7)[C]	161(1.7)[C]	569(1.5)[B]	1138(1.5)[A]	p<0.001
Al	401(1.6)[A]	301(2.2)[B]	399(1.8)[A]	233(1.8)[C]	p<0.001
P	141(1.2)[C]	140(1.3)[C]	184(1.3)[A]	159(1.3)[B]	p<0.001
K	807(2.2)[A]	561(2.1)[B]	671(2.4)[AB]	804(2.1)[A]	p<0.01
Ca	713(1.5)[D]	826(1.5)[C]	2535(1.5)[A]	2301(1.4)[B]	p<0.001
Mn	34.9(1.8)[B]	28.8(2.1)[C]	45.1(1.7)[A]	41.6(1.7)[A]	p<0.001
Fe	431(1.7)[B]	329(2.0)[C]	759(1.7)[A]	479(1.8)[B]	p<0.001
Cu	8.2(1.3)[D]	9.5(1.4)[C]	12.3(1.4)[B]	18.8(1.7)[A]	p<0.001
Zn	68(1.6)[C]	125(1.4)[B]	148(1.3)[A]	146(1.4)[A]	p<0.001
Sr	7.5(1.6)[C]	7.1(1.6)[C]	12.4(1.5)[B]	25.4(1.5)[A]	p<0.001

Mean with the same character (e.g. [A, B]) is not significantly different from each other by Duncan's multiple range test at p<0.05.

Sodium, K, Mn, and Sr (for both sexes) and Mg (for females) concentrations were higher in adults than in children.

In contrast, differences in mean hair elemental concentrations among the four villages were very dominant (Table 2); even among subgroups broken down by sex and age, the inter-village differences were significant except for Mn in female children. Sodium concentration in Dorogori was extremely high compared to the other three villages. Magnesium, Cu, and Sr concentrations were also the highest in Dorogori. Concentrations of these three elements in Rual and Wonie were low and those in Ume, intermediate. Potassium concentration was high in Rual and Dorogori. Calcium and Mn concentrations were higher in Ume and Dorogori than in Rual and Wonie. Regarding Zn concentration, the value was markedly low only in Rual. Phosphorus and Fe concentrations were the highest in Ume. Aluminum concentration was high in Ume and Rual, and low in Dorogori.

In principal components analysis for hair element concentrations (Hg concentration was involved in this analysis) in each of the four villages, factors with eigenvalues over 1.0 were extracted and varimax rotation was conducted. The factors extracted numbered three for Rual, four for Wonie and Dorogori, and five for Ume. The loadings of elements in factors 1 and 2 by village are plotted in Fig. 1. The comparison of the figures demonstrates similar component structure in three villages, except Dorogori; factor 1 had large positive loadings on Mg, Ca, and Sr, and factor 2 on Al, P, and Fe. In the case of Dorogori, factor 1 had large loadings on Al, P, K, Mn, and Fe, and factor 2 on Ca and Sr; factor 3 had large loadings on Na, Mg, and Cu, and factor 4 had negative loading on Zn and positive loading on Hg.

Figure 2 shows the profiles of average values of hair element concentrations, water element concentrations, and element intakes (from foods and drinking water) in the four villages. The profiles demonstrate that inter-village variations in hair element concentrations and in element intakes were relatively small, except for Na, Mg, and Ca in the former and Na, Al, and Fe in the latter. The profiles of water element concentrations, except for Al and Mn, markedly differed among four villages, in particular, between Dorogori/Ume villages and Wonie/Rual villages.

Both profiles of hair element concentrations and element intakes were surprisingly similar for the part from Ca to Sr. However, levels of P and K were relatively high in the element intake profiles but low in the hair concentration profiles, while Al level was relatively low in the former but high in the latter. The profiles of hair element concentrations were rather similar to those of water element concentrations in Na, Mg, Ca, Cu, and

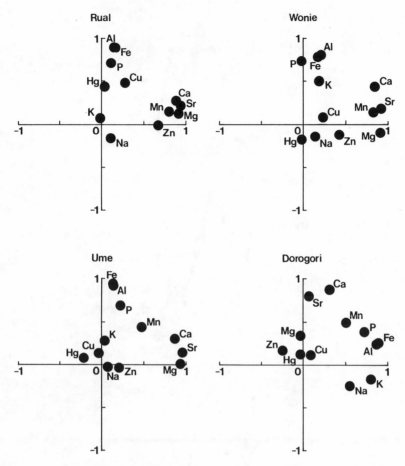

Fig. 1. Factor loadings of each element in factor 1 (horizontal axis) and factor 2 (vertical axis) for the four villages.

Sr. Magnesium, Cu, and Sr concentrations in hair and water were the highest in Dorogori, the second highest in Ume, and low in Wonie and Rual. In both hair and water Ca concentrations, Dorogori and Ume had high levels and Wonie and Rual, low levels. Hair and water Na concentrations were high in Dorogori and low in Wonie and Rual, although in Ume hair Na concentration was low despite the intermediate level in water Na concentration.

These relationships are ascertained in the scattergram between element intakes and hair element concentrations (Fig. 3). First, there exists a general trend that the larger the element intake the higher the hair ele-

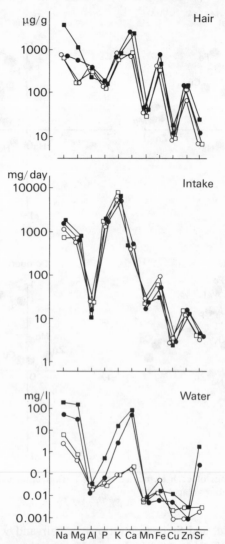

Fig. 2. Profiles of non-washed hair element concentrations, element intakes, and water element concentrations for the four villages.
○: Rual, □: Wonie, ●: Ume, ■: Dorogori.

ment concentration. The interrelation between element intakes and hair element concentrations using a number of elements has not been reported; the high correlation between the two levels in the present materials is

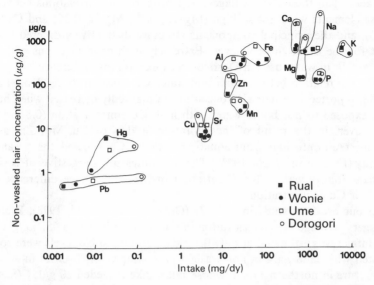

Fig. 3. Relationship between element intakes and non-washed hair element concentrations by village.

worthy to be thoroughly examined, comparing with other findings from various populations.

Second, Na, Mg, Ca, Sr, and Cu concentrations in hair vary to a considerable degree, irrespective of the intake levels. As has been mentioned, there is a high possibility that the inter-village differences of these element concentrations in hair were related to those in water as contamination sources.

From the study of Japanese immigrants in South America, Tsugane (1986) reported that hair Mg and Ca concentrations were well related to these element concentrations in well water, and that frequency of hair washing was positively correlated with the hair Mg and Ca concentrations. Another report indicated that Ca was deposited onto hair by washing with tap water (Marumo, 1983). On the other hand, significant positive correlations between any pairs of Mg, Ca, and Sr concentrations in hair were reported in various papers (Imahori *et al.*, 1979; Takeuchi *et al.*, 1982; Kamakura, 1983; Morita *et al.*, 1986; Moon *et al.*, 1988). In the case of the Gidra, correlation coefficients between these three elements were high (r = 0.71–0.90) in all the villages except Dorogori, and the principal components analysis showed large loadnings on Mg, Ca, and Sr in factor 1 in these three villages. In Dorogori, the correlations between these elements were not so high (Mg-Ca, 0.40; Mg-Sr, 0.36; and Ca-Sr, 0.65), and the principal components structure distinctively differed from that for the other three villages. Extremely high concentrations of Mg, Ca, and Sr in water used in Dorogori may explain this difference.

For the relation between hair and water Cu concentrations, Doi et al. (1988) reported that hair Cu concentration markedly increased when hair was exposed to hot boiler water with high Cu concentration (0.8 mg/l). Moreover, in the result of incorporation studies of Cu, Mn, Zn, and arsenic (As) onto hair from aquatic solutions, Cu showed the greatest affinity (Fergusson et al., 1983). These findings are suggestive of a significant role of water Cu concentration in the inter-village difference of the hair Cu concentration.

In our previous study in 1971–72 (Ohtsuka and Suzuki, 1978), Wonie villagers' average Na concentration in non-washed hair was 651 µg/g and Na intake was estimated at only 0.2 g/day, and these Na levels were contrasted with 1138 µg/g in non-washed hair samples of dwellers in Akita Prefecture in northern Japan, whose salt intake exceeded 20 g/day (Sasaki et al., 1981). In 1981, Na intake in Wonie has increased to 0.7 g/day Chapters 10 and 14 in this volume), but non-washed hair Na concentration (634 µg/g) has remained at the level similar to that in 1971–72. In Dorogori, Na intake was 1.8 g/day in 1981, but Na concentration in non-washed hair was 3750 µg/g, higher than that in Akita dwellers. These findings are suggestive of a significant role of water Na concentration in its hair concentration, even though the contribution of Na intake, particularly in the case of Dorogori whose Na intake was distinctively high, could not be neglected. In fact, when adsorption of Na to hair was tested by immersing hair samples in different concentrations of NaCl solution, Na penetrated into the porosity in hair and remained there after drying (Takemori, 1976).

NUTRITIONAL ASSESSMENT

Hair Mn level of the Gidra (after washing) was high (29.5 µg/g), compared to the values in the literature. Ross et al. (1986) attributed high Mn concentrations in detergent-washed hair of the Wosera people in East Sepik Province, Papua New Guinea (26.3 µg/g for male adults and 25.1 µg/g for female adults), to their high dietary Mn intake levels (7.5–11.0 mg/day); this Mn intake was lower than that of the Gidra (17.2–30.6 mg/day).

Australians (Aborigines) in the Gulf of Carpentaria, in which environmental Mn level was high, had high incidence of neurological disturbance

(Cawte, 1984; Kiloh *et al.*, 1980). This high incidence was related to their high Mn concentrations (34 μg/g for males and 77 μg/g for females) in hair (washed with detergent); these levels were markedly higher than those of Aborigines in other areas, i.e. 2.0 μg/g on the average (Stauber *et al.*, 1987). Hair Mn level in Caucasians in the same area (9.4 μg/g for males and 14 μg/g for females) was lower than that of Aborigines but was much higher than that in Caucasians living Sydney (0.6–1.2 μg/g). The same authors pointed out that the high hair Mn of Caucasians could be explained by the incorporation of Mn into hair from air dust dissolved in sweat, but the higher Mn level of Aborigines should come not only from exogenous source but from their high Mn-containing diet. In the Gidra with high Mn levels in intake and in hair, incidence of neurological disturbances should be carefully examined; in this course, their high intake of Fe, which competes with Mn for similar binding and absorption sites in the intestinal tract (Thomson *et al.*, 1971), should be considered.

Hair Al and Fe concentrations of the Gidra were high when the levels after washing were compared to the reported values; the non-washed levels were about 10 times higher than those in hair samples (non-washed) collected in Amazonas Indian Tribes in Venezuela (30.6 μg/g for Al, 65 g for Fe)(Perkons *et al.*, 1977). Intake levels of these elements among the Gidra were also high (Chapter 10 in this volume). However, the intervillage differences in the hair Al and Fe concentrations did not depend on those of their intake levels nor on those in concentrations in water. These discrepancies may be attributed to the different bioavailability of these elements by food group in connection with dietary food composition. At the same time, these discrepancies suggest great influence of contamination from soil. This is in agreement with Tsugane's (1986) finding that hair Al and Fe concentrations of Japanese immigrants in South America might reflect these element levels in soil. Although clinical influence of these element concentrations in hair has not been investigated, the markedly high levels in the Gidra hair should be related to intake and contamination levels on the one hand and physiological or nutritional significance on the other.

Hair Zn concentration in the Gidra was lower than the reported levels, in particular in Rual (68 μg/g on the average). It is interesting that their Zn intake per day (12.1–16.4 mg) was not markedly low compared to 15 mg of dietary intake recommended in the USA (National Research Council, 1980). Among children in an Iranian village, whose Zn intake was estimated at 19.0 mg/day (Maleki, 1973), Zn deficiency as shown in growth retardation was observed and was attributed to the high phytate content of their staple food, unleavened whole-wheat bread, which inhibits Zn absorption (Reinhold, 1971, 1972). Similarly, it is known that crude fiber inhibits Zn absorption (Solomons, 1982; Drews *et al.*, 1979). Also reported is the fact that high Fe/Zn molar ratio of more than 2.5, especially in case of nonheme Fe from plant foods, inhibits the intestinal uptake of Zn (Solomons and Jacob, 1981). In this connection, the low Zn status of the Wosera in Papua New Guinea was attributed to their dependency on plant foods and their high Fe/Zn ratio (Ross *et al.*, 1986). In the case of the Gidra, high crude fiber intake, i.e. 10.0–16.8 g/day (Chapter 9 in this volume), is judged as the major reason in decreasing hair Zn level. The particularly low hair Zn level in Rual is explained by the facts that Zn intake was lowest and Fe intake was highest, and thus Fe/Zn molar ratio was extremely high (9.4) in this village.

Strain *et al.* (1966) reported low hair Zn concentrations in Zn-deficient male Egyptian dwarfs (54.1 ± 5.5 μg/g) compared to normal Egyptians (103.3 ± 4.4 μg/g), and oral zinc sulfate therapy increased average hair Zn level in the dwarfs to 121.1 ± 4.8 μg/g with clinical alleviation of the deficiency syndrome. Low hair Zn levels were also reported in children with a similar syndrome in a study of an Iranian village (Eminians *et al.*, 1967). In these reports, it was considered that hair Zn concentration of less than 70 μg/g represented marginal Zn deficiency. If this criterion is accepted, Zn status in 13 % of the Gidra (51 % of the Rual villagers) falls in the marginally deficient category. In children, a relationship between low hair Zn levels and retarded growth and hypogeusia has been reported (Hambidge *et al.*, 1972, 1976; Buzina *et al.*, 1980; Chase *et al.*, 1980; Gentile *et al.*, 1981; Xue-Cun *et al.*, 1985; Thompson *et al.*, 1986; Vanderkooy and Gibson, 1987). These studies suggested that hair Zn level could be used as an indicator of long-term Zn nutriture because both retarded growth and hypogeusia occurred as a result of continuous inadequate Zn intake. The relationship between hair Zn level and the growth of children in the Gidra will be reported in our forthcoming paper.

Another feature—that hair Zn concentrations in the Gidra was lower in adult females (106 μg/g on the average) than in adult males (135 μg/g)—should be noted, since the importance of adequate maternal Zn nutriture for reproduction has been demonstrated (Hambidge *et al.*, 1986). Furthermore, the Zn concentration of pregnant or lactating women (89 μg/g) was significantly lower than that of the other adult women (114 μg/g); in Rual, this concentration in 12 pregnant/lactating women was markedly low (46 μg/g) compared to the other women's value (67 μg/g). On the basis of measurement of hair collected from Wonie in 1971–72, the Zn status of female villagers was judged to be marginally deficient; of seven births during the survey period, three stillbirths and one infant death three months after birth were recorded (Ohtsuka and Suzuki, 1978). After a ten-year interval, hair Zn concentration of Wonie females has increased to 123 μg/g. Judging from the low hair Zn level in Rual females (lower than males), severer Zn deficiency may have continued in this village, though no stillbirths were observed in Rual during five months in the 1981–82 survey period. The fact that reproduction rate of women in northern villages of the Gidra, including Rual, was particularly low in the past (Chapter 18 in this volume) may be related to the Zinc nutriture. Thus, long-term follow-up studies are necessary to confirm the relationship between the low hair Zn levels and health problems in the Gidra.

(Tetsuro Hongo and Tsuguyoshi Suzuki)

Chapter 17

Mercury Intake and Hair Mercury

Kyle (1981) and Kyle and Ghani (1982) reported high mercury con-centraions in hair of the people inhabiting the Lake Murray area in the Fly and Strickland River catchment of Papua New Guinea. Two problems are of special importance in human ecology and in primary health care. First, the Fly and Strickland area is judged to be free from anthropogenetic mercury pollution, although the geochemical reasoning is not clear. Second, the hair total mercury level of the Lake Murray people (mean value of 17.9 $\mu g/g$, highest value of 58.4 $\mu g/g$) is one of the highest in the world, in populations not exposed to artificial mercury pollution. The subjects of their study live in a riverine or lake-side environ-ment and frequently eat fish, which is the main mercury source. Partic-ularly high mercury concentrations of over 0.5$\mu g/g$ were detected in the muscle of barramundi perch or giant perch (*Lates calcarifer*) caught in the Fly River system in 1979 (Kyle, 1981; Sorentino, 1979). The hair methyl mercury concentration in the Lake Murray inhabitants, who ate this fish two to three times per day, was estimated at 15.5 $\mu g/g$ on average (Kyle, 1981; Kyle and Ghani, 1982).

The Gidra people still basically subsist on local foods, and the amount of fish consumed is determined mainly by the village locality (see Chapter 9 in this volume). In fact, villagers living inland eat negligible amounts of fish, and our measurement of scalp hair samples of the Wonie villagers collected in 1971 revealed that the total mercury level was only 1.8 $\mu g/g$ for men and 1.4 $\mu g/g$ for women (Ohtsuka and Suzuki, 1978). However, among Ume and Dorogori villagers, fish and other aquatic resources are the major animal foods.

The ecological setting of the Gidra is suitable for the comparative analy-sis of hair mercury levels in the diversified habitat. In our field research in 1980 and 1981, four villages, i.e. Rual, Wonie, Ume, and Dorogori, which differ ecologically from each other, were selected for the intensive survey. Hair mercury concentrations of 289 people from these four vil-lages were measured. Their hair mercury levels are also discussed in rela-tion to our chemical analysis of mercury concentration of their staple

Table 1. Number of Hair Samples Analyzed

Sex	Age-grade[a]	Village			
		Rual	Wonie	Ume	Dorogori
Male	*Miid/Nanyuruga*	3	2	9	3
	Rugajog	10	10	11	12
	Kewalbuga	4	8	10	5
	Yambuga	3	5	10	10
	Sobijogbuga	8	6	10	7
Female	*Nanyukonga*	6	4	10	9
	Kongajog	10	10	10	11
	Ngamugaibuga	10	10	10	5
	Sobijogngamugai	9	10	10	9

[a] Estimated ages of each age-grade are shown in the frontispiece.

foods and the estimated mercury intake from foods, based on our food consumption survey.

METHODS

As has been described in the previous chapters, animal and plant food samples (Chapters 7 and 8) and hair samples (Chapter 16) were collected in the four Gidra villages.

The number of hair samples analyzed are shown in Table 1. The samples were stored in a desiccator. Prior to mercury measurement they were dried at 85°C for 4 hr to minimize variation due to the different water content of the hair. (Using this drying method the loss of mercury from the hair sample was negligible).

Mercury was measured using the modified Magos method (Magos, 1971; Yamamoto *et al.*, 1980). When organic and inorganic mercury were measured selectively, the sample of hair or food was cut into pieces or mashed, and then extracted twice with 1 N HCl solution. In this process, most of the organic mercury was removed and the inorganic mercury remaining in the residue was measured without interruption due to breakdown of coexisting methyl mercury. To guarantee the accuracy of the measurement, the reference hair specimens (certified mercury value: 4.4 μg Hg/g) provided by the National Institute of Environmental Studies of Japan were measured each time.

To estimate the amount of mercury intake, this study used the results of our food consumption survey for 14 days in four villages in the dry season of 1981 (Chapter 9 in this volume). It was questionable, however, whether the above procedure reflected adequately the general level of metal intake; in particular, mercury was concentrated in specific kinds of fish, and catch of each fish species was markedly sporadic in the area.

RESULTS

Mercury Level in Foods

Plant foods contained small amounts of total mercury, usually below 0.01 μg/g; the highest concentration was 0.026 μg/g in galip seeds (Fig. 1; also see Appendix). Among animal foods, fish showed the highest level of total mercury, varying from 0.038 μg/g to 1.86 μg/g, with the inorganic to total mercury ratio being several to 23%. Reptiles had the next highest mercury levels. Except for turtles, the level of total mercury was over 0.1 μg/g (the ratio of inorganic to total mercury: 10% to 16%), and the flesh of water snakes contained 1.3 μg/g mercury. In other animal foods, moderately high levels such as 0.1 μg/g or 0.2 μg/g were found in shellfish and crustaceans. Land mammals, birds, and insects had lower levels of total mercury than fishes and reptiles but slightly higher levels than plant foods.

Estimated Mercury Intake

From the results of the household food consumption survey, fish consumption was the greatest in Dorogori, the coastal village, next in Rual, the northern village, then in Ume, the riverine village. In Wonie, the inland village, no fish was eaten. The fish eaten most often in the coastal village were sharks, catfish, threadfin salmon, barramundi perch, Jardine's barramundi, and imported canned mackerel. Catfish, gudgeon, canned

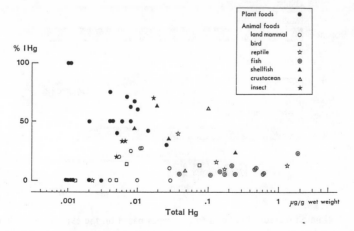

Fig. 1. Mercury contents of the Gidra foods.

mackerel, and other unidentified freshwater fish were eaten in Rual, while Jardine's barramundi, barramundi perch, canned mackerel, and other unidentified freshwater fishes were eaten in the Ume. Few reptiles (as compared to fish) were eaten in any of the villages. In particular, the sea turtle eaten in Dorogori contained a low level of mercury (0.002 μg/g).

From the mercury values of foods and the food consumption records, total and organic mercury intake were estimated (Fig. 2). In this estimation, mercury taken from imported foods and some local foods such as cassava, pumpkin, coconut, papaya, pineapple, winged beans, wallaby liver, and flesh of the flying fox was not included because the mercury content was not measured. For the unidentified fish that were not sampled for mercury analysis, the geometric mean of mercury values in freshwater fish was applied. The highest estimate, 81.3 μg/day/adult male for total mercury and 70.4 μg/day/adult male for organic mercury in Dorogori, seems to reflect the largest amount of fish consumption, mainly marine fish. In Wonie, where no fish was consumed during the survey period, the smallest value, 8.0 μg/day/adult male for total mercury was estimated. Several μg were added to the estimate of mercury in the shellfish and crustaceans eaten in Ume village.

The methyl mercury content of canned mackerel was not estimated, but it may be sizable because the chemical analysis of similar samples from Port Moresby, the capital of Papua New Guinea, detected a mean mercury

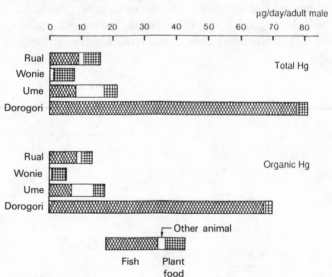

Fig. 2. Estimated mercury intake in four villages based on the 1981 household food consumption survey.

level of 0.17±0.11 μg/g for total (organic) mercury (Kyle and Ghani, 1983). Thus, the estimated amount of mercury intake would be increased by 4.2±2.7 μg/day/adult male for total (organic) mercury in Dorogori, and negligible in Wonie and Rual villages.

Hair Mercury Levels

The breakdown of total mercury concentrations in the hair of the Gidra subjects by village and sex/age group are shown in Fig. 3. The lowest levels (mean of 1.5 μg/g for all males; 1.0 μg/g for all females) were obtained for Wonie villagers, and this hair mercury level was very similar to the level found in our samples collected in 1971 in this village (Ohtsuka and Suzuki, 1978). This result coincides with the fact that it is very rare for the Wonie villagers to consume fish and other aquatic animals. Compared to the mercury level of Wonie villagers, the remaining village groups showed higher levels; the mean levels for all males and females were, respectively, 3.8 μg/g and 3.4 μg/g in Ume, 4.1 μg/g and 4.4 μg/g in Dorogori, and 7.1 μg/g and 6.4 μg/g in Rual. The difference among these three villages does not parallel that found in the estimated mercury intake.

The mercury level in females was significantly higher in Dorogori than in Ume. No significant difference was found in males.

To confirm that the difference in total mercury levels was due to different organic mercury levels, hair samples of ten adult males from each village were selected randomly to measure inorganic mercury content.

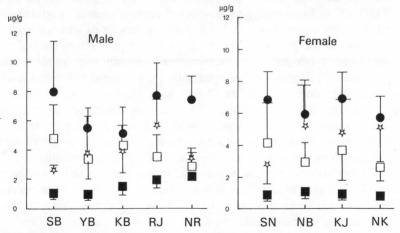

NR: *nanyuruga*, RJ: *rugajog*, KB: *kewalbuga*, YB: *yambuga*, SB: *sobijogbuga*, NK: *nanyukonga*, KJ: *kongajog*, NB: *ngamugaibuga*, SN: *sobijobngamugai*.

Fig. 3. Hair total mercury levels by village and sex/age group. ●: Rual, ■: Wonie, □: Ume, ☆: Dorogori.

Table 2. Proportion of Inorganic Mercury to Total Mercury in Hair of Adult Males (mean±SD)

Village	N	Total Hg (μg/g)	Inorganic Hg (μg/g)	% Inorganic Hg
Rual	10	7.7 ± 2.2^A	0.73 ± 0.23^B	9.8 ± 2.2
Wonie	10	1.9 ± 0.5^B	0.16 ± 0.05^B	8.2 ± 1.2
Ume	10	3.6 ± 1.4^B	0.35 ± 0.11^B	10.2 ± 2.0
Dorogori	10	5.8 ± 2.0^A	0.66 ± 0.21^A	12.0 ± 2.9
ANOVA		$F = 22.21, p < 0.01$	$F = 26.55, p < 0.01$	$F = 5.26, p < 0.01$

Means with the same character ($^{A, B}$) do not significantly differ from each other by Duncan's multiple range test at $p < 0.05$.

As a result, the proportion of inorganic to total mercury was approximately 10% (Table 2), and the inter-village difference was minimal.

DISCUSSION

Of all foods, fish in Papua New Guinea and in other parts of the world have among the highest mercury concentrations (Piotrowski and Inskip, 1981). According to the Environmental Health Criteria Document issued by the World Health Organization (WHO), mercury concentrations in freshwater fish from uncontaminated water were between 0.1 μg/g and 0.2 μg/g (wet weight). Most oceanic fish have levels of about 0.15 μg/g, although the large carnivorous species fall in the range of 0.2–1.5 μg/g (WHO, 197 6). In this study, several species such as Jardine's barramundi, and threadfin salmon had clearly elevated mercury levels when compared with those in the WHO Document. As for the high mercury level of the barramundi perch, mercury concentration from only one sample in this study was 0.15 μg/g, which was markedly lower than the reported levels (Kyle, 1981; Sorentino, 1979). Sharks have been well known to have high mercury levels in the muscle. In Australia, several species were reported to have levels over 0.5 μg/g (Working Group on Mercury in Fish, 1980). The sharks caught in Dorogori were ichthyologically unidentifiable, but the high mercury level was in the range of the literature values.

Mercury concentration in reptiles other than the sea turtle was comparable with that in fish. This is reasonable since most reptiles occupy the higher trophic level. Reptiles have been neglected as a source of mercury, because they are rarely eaten. However, the high mercury levels in the present results suggest that the reptiles' contribution has to be evaluated in anthropological populations who eat them regularly or frequently. Mercury concentrations in food other than fish are below 0.06 μg/g according to the WHO Document (WHO, 1976). This was also the case in the present food of the Gidra; the exceptions were reptiles and some shellfish and crustaceans.

The present estimate of mercury intake based on food consumption survey data was not consistent with the average level of mercury in hair. The highest level of hair mercury was found in Rual village, where the estimated mercury intake was the third highest among the four villages. In Dorogori, the coastal village, where the estimated mercury intake was extraodinarily high at 80 μg/day/adult male, the hair mercury levels were lower than those in Rual village and comparable with that in Ume, the riverine village. There are several possible reasons for this inconsistency. First, in Rual, the villagers sometimes stay in their fishing camp and the food consumption survey did not cover the presumably large fish consumption during this time. Second, though sharks were rarely caught in Dorogori, i.e. only once during the seven months that our research team member was there, the large shark had provided ample amounts of flesh and supplied mercury at a level of about 50 μg/day/adult male during the food consumption survey period. Moreover, the catch of fish in this village is generally unstable, and an unusually large catch might have occurred in the survey period. Third, hair sampling was conducted not on a single occasion but whenever villagers cut their hair. Therefore, the time of sprouting of the sampled hair did not necessarily coincide with that of the surveyed food consumption, and the average values of hair mercury level must have been approximate estimates covering a few months.

Residents of the four Gidra villages had health checkups in 1981. The results of anthropometry, urinalysis, blood pressure, and visual acuity as well as demographic analysis (overall mortality and fertility) have been reported in other chapters of this volume. Although an exact neurological examination was not conducted, no apparent signs and symptoms of suspected methyl mercury poisoning were noticed by a physician in our team. Considering that according to WHO's report (WHO, 1980), 10 μg/g to 70 μg/g hair mercury levels in pregnant women and children may be related to elevated frequency of mercury-induced abnormalities, it is noticeable that eight out of the 289 Gidra subjects had concentrations above 10 μg/g. In comparison with the Lake Murray people in the upper Fly and Strickland area, the Gidra hair mercury level is low, but could still possibly cause abnormalities. Since the Gidra people do not depend heavily on fish for their animal food, more surveys are needed to study the mercury level in human populations, whether they are subjected to mercury pollution or not.

(*Tsuguyoshi Suzuki*)

IV. POPULATION STRUCTURE AND DYNAMICS

Overview

Population structure and dynamics, represented by fertility, mortality, reproduction, and increase rates and migration rates, are basic aspects in population ecology. In our view, it is important to identify the interrelations between these demographic parameters and the people's adaptive mechanisms in the senses of, for instance, subsistence and nutrition. Microdemographic techniques have recently advanced the point where they can be used to analyze demographic phenomena in small groups. Even so, most of the newly developed techniques presuppose that ages of the subject people are given. For the so-called anthropological populations including the Gidra, whose ages cannot be determined, we must use alternative techniques.

Our efforts to study the Gidra demography were devoted primarily to collecting reliable data by repeated interviews. The rate of population increase in the past is indispensable for revealing long-term adaptive mechanisms and for fully understanding the current adaptive situation which we can observe. For this purpose, the rate of inter-generational replacement of females, derived from the Gidra genealogies, are applied to a formula to estimate annual increase rates; the results are analyzed not only for the historical trend among the Gidra as a whole but for the differences among sub-populations in relation to local adaptive mechanisms (Chapter 18). Using the genealogical data and the interview data for detailed reproduction histories for once-married females, the recent fertility and mortality transitions are also estimated (Chapter 19). Furthermore, this study points out some analytical problems with demographic parameters when using the records from contemporary people.

More than half of the Gidra people's marriages have been organized within the same village; in the bulk of the remainder, spouses from different Gidra villages (mostly the brides) moved to the other partner's village, judged from the living members' migration histories (Chapter 20). In the same chapter, the inter- and intra-population migrations of the current members demonstrate the demographic infrastructure of the Gidra population on the one hand, and on the other the gradual change from a closed system to an open system with an increasing number of out-migrants.

Chapter 18

Population Growth in the Past

Despite the importance of understanding human adaptation and evolution, it is difficult to know the degree to which a nonliterate group has increased its population in the period when it is little influenced by civilization. Three different ways of approaching the problem have been prevalent: 1) palaeo-demographic analysis of skeletal and other archaeological remains, 2) analysis of information collected in contemporary populations, and 3) microsimulation analysis. Each of these methods has its own defects. Palaeo-demography contributed to comprehension of long-term change, with crude accuracy. The study of contemporary populations may provide data, though fragmentary, suitable to demographic analysis. For example, Neel *et al.* (1977) recommend application of model life tables in using the population growth of tribal groups as an indicator for modeling human evolution. Nonetheless, there is a good possibility that the demographic traits of a population being studied have been modified, to some extent, by the influence of civilization. A simulation analysis cannot fully exert its advantages without a set of appropriate parameters derived from empirical data.

We attempted to derive the population increase rate of the Gidra over approximately the last century based on their genealogical records, involving 889 married women who completed reproduction. The indicator used was how many daughters of each married woman survived to marry. The basic principle of this method is net reproduction rate, which measures the inter-generational replacement of females and provides the rate of natural population increase (Pressat, 1972).

As was pointed out by Bayliss-Smith (1977), in Melanesia there exist small human populations that operate within well-defined boundaries of marriages. One of the most relevant reasons comes from the linguistic diversity of Melanesians, especially of the speakers of Papuan (Non-Austronesian) languages; Wurm (1982) estimates the total number of Papuan languages at about 750 and, among approximately three million speakers, the number of per-language speakers at only 4000. The Gidra are a typical example of such populations. In fact, even at the present time (1980) the rate of migrants from other populations was only 0.032 among all members and only 0.054 among the married members. Also noted is that the Gidra had been to a negligible degree influenced by civilization before WWII. For instance, Christianity, primary school education, and local government were introduced to their society only in the 1960s; on the other hand, the Gidra's out-migration to urban areas has prevailed since the 1950s or 1960s in the coastal and riverine villages and since the 1970s in the inland villages (Chapter 20 in this volume). These facts imply that the Gidra can be treated as an entity of survival through generations, at least before several decades ago (cf. Birdsell, 1973). An associated question is intra-population variation in the rate of population replacement, when the women are broken down by characteristics such as birthplace.

MATERIALS

Up until several decades ago, Gidra inhabited loosely organized settlements, although any place where Gidra members, either present or past, lived is easily recognized by the contemporary members as a main or satellite settlement of one of the 13 contemporary villages. Although their villages are relocated every few years or every few decades even in the present time, such relocations were not accompanied by radical change in environmental conditions, except in one case in the 1920s when the Dorogori moved from the bank of the Oriomo River (some 10 km upstream from the mouth) to the present site, i.e. the coast facing Daru Island, the capital of Western Province.

Regarding marriage customs, exogamy between two moieties, each involving about 20 clans, is still observed to a considerable degree. Traditionally, it has been ideal for two marriages to be organized simultaneously by means of an exchange of women. Among the married members of Wonie in 1971–1972, 70% followed this system (Ohtsuka, 1983); in most independently organized marriages, there were special reasons such as death of the wife soon after the marriage or cases of delayed compensation for unbalanced relations between the clans. These customs have made it important for the villager to memorize the marital relations of his/her ascendants. In other words, the ascendants, even those deceased, tend to be well remembered if they survived to marry.

There are three more customs directly relevant to this study. First, it is a rule that a female marries soon after menarche. According to my estimation from several kinds of evidence such as patrol officers' records, the birth order of the villagers, and a limited number of exact dates (in terms of year) of their births, the age of female marriage is 17–19 years.

Because of a wider range of male age at marriage and the relative predominance of polygynous marriages, this cultural norm has been observed to a considerable degree, even if the recent modernization tends to raise this age. Second, when a woman becomes a widow owing either to her spouse's death or to divorce, she will usually remarry if she has not passed reproductive age. Third, the Gidra informed us that they have seldom practiced birth control or infanticide. In fact, we have encountered no episode of these practices at least in the last half-century when the present members have lived, even though their traditional knowledge includes several methods of contraception, artificial abortion, and infanticide, as reported by my co-workers (Akimichi, 1984; Suzuki, 1985). In our understanding, this may be related to a prohibition on marriage within the moiety and relatively unrestricted premarital sexual relations (Suzuki et al., 1984a). Pregnancy of unmarried females occasionally triggers a marriage between the female in question and her partner, whether married or unmarried. Alternatively, such offspring are reared by the unmarried mother and later taken to a household in which she becomes a wife.

It can thus be concluded that the Gidra have tended to have as many children as possible. This assertion is partly supported by the facts that most severe quarrels between a married couple concern the wife's nonpregnancy or lack of offspring (Ohtsuka, 1983) and that men who have had many children, mostly by means of polygynous marriages, have been admired by the members of their own clan (the social organization of the Gidra is patrilineal).

During our two field investigations, for three months in 1980 and for six months in 1981–1982, genealogical data were collected in all Gidra villages by means of consistent interviews with the villagers in question and with aged villagers in particular; most time in the latter investigations was devoted to revising the data collected in the former. In this study, name, sex, clan, birthplace and living or death place, and marital, parent-child, and sibling relations (in biological terms) of all individuals, alive or dead, were identified. The deceased villagers who were listed included those two to four generations back from the contemporary members at reproductive age. In this process, special attention was paid to involvement of all deceased members who had no surviving descendants; questions about them were asked for several aged informants in each village. The other set of data relevant to this study was detailed reproductive histories of living women in four villages (Rual, Wonie, Ume, and Dorogori), in each of which one of four research team members including myself was stationed for several months.

METHODS

Except in genealogical-demographic studies for populations with historical records such as church registers (e.g. Skolnick· et al., 1976; Brennan and Boyce, 1980; Workman and Devor, 1980; Brennan, 1981), genealogical information has seldom been used for demographic analysis. For instance, a half-century ago, Powdermaker (1931) analyzed genealogical data on living people in five villages of New Ireland (Melanesia). The present study is similar to Powdermaker's in the basic idea but treats not only living members but also deceased members.

The indicator applied to measuring inter-generational replacement of females in this study was the average number of daughters who were born to a group of married women (as mentioned below, classified according to the estimated generations) during their lifetimes and who survived to marry. Unmarried women are not involved in this analysis, although there are very few cases of women who survive to marital age but do not actually marry. For convenience, this indicator is called daughter-mother ratio, or DMR, in this analysis. Theoretically, the application of DMR is valid under the supposition of a fixed pattern of age-specific fertility and mortality rates, as in the application of net reproduction rate. Also necessary is a supposition that the age of female marriage has not changed over generations. It is impossible to determine that these suppositions are applicable to the Gidra data, in which even ages are not known. In my judgment, however, the fertility and motality pattern and age at marriage did not differ among generations so markedly as to reject the above suppositions, since the bulk of childbearings of the subject women occurred before 1945, or in the period of less influence of civilization. With some reservations regarding this methodology, the present analysis will treat DMR as a substitute of net reproduction rate.

The rate of natural increase, or Lotka rate (P), is calculated from the net reproduction rate (R_0) as follows:

$$P = {}^y\sqrt{R_0} - 1,$$

where y is the mean age at childbearing (Pressat, 1972); Hassan (1981) estimates y in prehistoric populations at approximately 20 years. The Lotka rate equals the rate of annual population growth (r) calculated by the following equation:

$$r = \frac{1}{t} \cdot \ln\left(\frac{N_2}{N_1}\right),$$

where N_2 is the population size reached from an initial population, N_1, after a time period (t) (Hassan, 1981).

The subjects of this study were females who had married and died or, if still living, had passed reproductive age. Judgment as to the completion of reproductive age for the living women was based on two criteria. The first was the lapse of approximately seven years or more after delivery of the last child, the last child having passed the first grade of the Gidra's traditional age-grade system (see frontispiece). The second criterion, used for sterile and seldom-pregnant women, was to estimate the age of the woman in question through a comparison with her siblings and/or other consanguines. Among the deceased women, those whose parents and siblings (who had survived to marry) could not be ascertained were excluded from the analysis for the sake of accuracy.

The next process was to categorize the subject women into the following four groups, taking their generations into account: group A, not all female offspring have married; group B, all female offspring have married; group C, all female offspring have completed reproduction; and group D, all female offspring of all female offspring (or all granddaughters of the subject women through the female line) have married. In this process, the grouping of sterile or less-pregnant women was done by means of a comparison like that used for the judgment of completion of childbearing. It is natural to assume that the age interval between the groups is approximately 20 years. It is also assumed that, the older the generation, the more this categorization deviates from the true situation. The numbers of subject women, classified as alive or dead at the time of the survey, are given in Fig. 1, in which their assumed birth years are also illustrated. It can be judged, however, that the overlap of age ranges between the groups should have produced no serious biases in the results.

RESULTS

Table 1 shows the number of children who were born to women of each group and survived to marry, and the corresponding DMR and Lotka rates. For calculating the Lotka rate, the mean age of childbearing

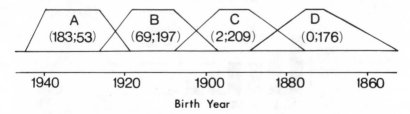

Fig. 1. Assumed birth years of the subject women. Numbers of living women (on the left) and deceased women (on the right) are shown in parentheses.

Table 1. Daughter-Mother Ratio (DMR) and Lotka Rate (P) by Mother's Group

| | No. of married children | | | P at mean childbearing age | |
Group (N)	Male	Female	DMR	20	25
D (176)	186	184	1.0455	0.0022	0.0018
C (211)	228	217	1.0284	0.0014	0.0011
B (266)	283	282	1.0602	0.0029	0.0023

(y in the above-mentioned equation) was taken at two levels, 20 and 25; according to my estimation, the mean age for the Gidra was within this range. With slight inter-group variations in the DMR and Lotka rate, an overall characterization highlights their low values.

The women of each group were classified according to number of married children of both sexes (Fig. 2). The three figures prove, in particular, two observations. First, the distribution pattern of group D markedly differs from that of group C or B, the former involving a much smaller proportion of women who had no married child. This seems largely attributable to the errors of the contemporary villagers' memory; the adop-

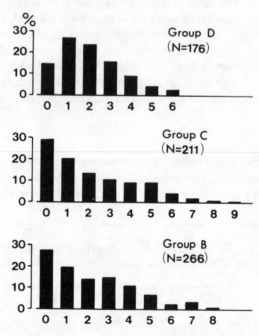

Fig. 2. Percent distribution of women by number of children of both sexes who survived to marry.

tions of children born to the mothers of group D occurred so long ago that the adoptive mothers tend to be recognized as the real mothers (adoption of a child, especially between a couple with no or few children and another with many children, has been popular in Gidra society [Ohtsuka, 1983]). It is thus judged that the distribution pattern of group C or B is the norm, and this leads to the second observation, that the proportion of women with no married child was nearly 30%, implying that the genes of these women were not transmitted to the following generation and also suggesting a high proportion of sterile women.

The reproductive histories obtained from thorough interviews with all living, ever-married women in four villages were analyzed mainly for examination of the reliability of the genealogical data, especially DMR. Of the interviewed women, 71 had passed reproductive age, 21 belonging to group B and 50 to group A. As to the genealogical data, women of groups C and B were divided into two sub-groups based on whether they were alive or dead at the time of the investigation, whereas women of group D were excluded from this analysis. Table 2 shows the percent distribution of women by number of live births, the per-woman number of live births (completed fertility), and, for groups C and B, the per-woman number of married daughters, equivalent to DMR.

The inter-group comparisons demonstrate several things. First, "genealogical B: living" and "interviewed B" are similar in the per-woman number of live births, the per-woman number of married daughters, and the distribution by live births, although, because of the small size of the sample, the distributions tend to fluctuate. Similarly, the distribution by live births and the per-woman number of live births are fairly identical between "genealogical A: living" and "interviewed A."

Second, the differences between "died" and "living" in either "genealogical B" or "genealogical A" are prominent, and two reasons seem to be relevant. One is that "died" should have involved a higher proportion of women who died during their reproductive years, and the other is that children who were born to mothers of "died" and who died in infancy or childhood had occasionally faded from the people's memory.

Third, there is a marked difference in the per-woman number of live births between "B" and "A" in both "genealogical: living" and "interviewed." This is judged to be due to such effects of modernization as nutritional improvement (Chapter 9 in this volume) and increased use of medical services (Akimichi, 1984); according to Hashimoto (1983), the town of Daru, the only center of modernization in this area, markedly developed after WWII. The central period of childbearing of mothers of group A was approximately between 1940 and 1970.

Fourth, despite the inter-group differences in the per-woman number

Table 2. Comparison of Genealogical Data with Data from Interview Survey

Group (N)	% Distribution of women by number of live births											Per-woman number of:	
	0	1	2	3	4	5	6	7	8	9	10+	Live births[a]	Married daughters[b]
Genealogical													
C: died (209) and living (2)	24.2	19.1	12.3	10.4	9.0	9.5	7.1	4.7	2.4	3.3	0.5	2.863	1.028
B: died (197)	22.8	17.8	8.1	11.7	9.6	10.2	7.6	5.6	3.6	2.0	1.0	3.010	1.020
B: living (69)	15.9	17.4	10.1	11.6	11.6	11.6	8.7	5.8	1.4	2.9	2.9	3.362	1.174
A: died (53)	24.5	22.6	11.3	3.8	11.3	9.4	5.7	3.8	0	3.8	3.8	2.887	–
A: living (183)	14.8	12.6	10.4	6.0	6.0	7.1	9.3	12.0	8.7	5.5	8.2	4.546	–
Interviewed													
B (21) and A (50)	16.9	9.9	12.7	5.6	5.6	5.6	12.7	8.5	8.5	9.9	4.2	4.366	–
B only (21)	19.0	9.5	9.5	9.5	14.3	9.5	9.5	4.8	9.5	4.8	0	3.714	1.238
A only (50)	16.0	10.0	14.0	4.0	2.0	4.0	14.0	10.0	8.0	12.0	6.0	4.640	–

[a] Per-woman number of live births equals completed fertility.
[b] Per-woman number of married daughters equals DMR.

of live births, the proportion of women with no live birth was similar among "genealogical B: living," "genealogical A: living," and "interviewed" at 15–17%; this percentage is recognized as the sterility rate of the Gidra women.

Another interest is the difference in DMR when the mothers are classified by marital or ecological conditions. The first and second conditions were mother's birthplace and death or living place. Taking the ecological (environmental and cultural) conditions into account, the places were categorized into 1) northern villages, Rual and Kapal, close to the river and the remotest from Daru; 2) inland villages, Iamega, Wipim, Podare, Gamaeve, Wonie, and Kuru, located on relatively higher land and apart from the river; 3) riverine villages, Ume, Wuroi, Woigi, and Abam; and 4) a coastal village, Dorogori. The third condition was whether the mother transferred from one village to another or not, determined by her birthplace and death or living place; this is treated as a crude indicator of woman's relocation at marriage because most inter-village migrations occur at marriage in the Gidra society. The last condition was whether the woman married once or twice or more.

The results are shown in Table 3, in which, owing to the small sample size, women of groups D, C, and B were combined. The DMRs broken down by the mother's birthplace markedly vary. The higher DMR of mothers born in the coastal village (Dorogori) may have related to the greater influence of modernization. More interesting is the fact that the

Table 3. Daughter-Mother Ratio (DMR) by Mother's Characteristics

Birthplace[a] (N)	Northern (82)	0.732
	Inland (332)	1.160
	Riverine (183)	0.891
	Coastal (22)	1.409
Death or living place[b] (N)	Northern (85)	0.976
	Inland (319)	1.110
	Riverine (198)	0.899
	Coastal (35)	1.514
Inter-village migration[c] (N)	Non-migrant (425)	1.009
	Migrant (228)	1.114
Number of marriages (N)	One (457)	0.967
	Two or more (196)	1.230

[a] Excluded are 34 mothers who were born outside the Gidraland and whose married daughters numbered 44.

[b] Excluded are 16 mothers who died or live outside the Gidraland and whose married daughters numbered 15.

[c] Including 34 mothers who migrated from and 16 mothers who migrated to outside the Gidraland.

DMR of mothers born in inland villages was higher than that of those in northern villages or riverine villages. The DMRs of the latter two were less than 1.0, implying decrease of population. A comparison by death or living place proved the reduced difference of DMR among northern, inland, and riverine groups. This is judged to be partly related to the difference in DMR broken down by inter-village migration. The difference in DMR between one-marriage women and plural-marriage women can be considered to be caused not only by the different numbers of marriages *per se* but also by the difference in duration of women's reproductive lifetimes, simply because the longer they survive the higher the chance of remarriage.

RELIABILITY OF GENEALOGICAL RECORDS

The discussion should begin with the reliability of DMR obtained from the genealogical records on which the present study was based. As demonstrated in Table 2, the reliability of DMR (or per-woman number of married daughters) for living women was assured by comparison with data from thorough interviews with all women in four selected villages. The DMRs for deceased women of group B are judged also to be reliable, since the present senior members belong to the same generation. Consequently, little doubt remains about the DMR for groups A and B.

The DMRs of three groups, D, C, and B, ranged from 1.0284 to 1.0602; taking into account the inconsistent trend on the one hand and the small number of subjects on the other, it is reasonable to judge that the DMR in that period was rather constant, fluctuating within a small range. The central reproductive years of these women were between 1880 and 1950. Since, as mentioned previously, the Gidra had scarcely been influenced by civilization before 1950, the similar level of DMR among the three groups was very plausible.

The above discussion suggests that the present genealogical-demographic method using DMR is valid for estimating the rate of inter-generational replacement or natural increase among a human population consisting of a small number of members, forming a unit in which most marriages are organized and (if not to the same degree of necessity) possessing a marriage custom like an exchange of women.

LOW RATE OF POPULATION INCREASE

When we suppose that the DMRs for groups D, C, and B deviated from the constant value and that the mean age of childbearing was 22.5, the Lotka rate or annual increase rate of the Gidra before 1945 was calculated at 0.0020 (when the mean childbearing age was taken at 20 and 25, the rate differed to a small extent, that is, 0.0022 and 0.0018, respectively). Whether the 0.002 increase rate is regarded as high or low is opportunistic, but at least we can say that this rate is, in general, lower than those reported from ethnographic studies in nonindustrial societies.

Because of the sparse data on the population increase rate, however, it is convenient to compare the per-woman number of live births; in the Gidra, this value was about 3.5 for the living women of group B and about 4.5 for the living women of group A, both of which should have been much higher than those for groups D, C, and B (including deceased members as well). Compiling ethnographic data, Nag (1962) reports that of 46 societies examined 70% show values of more than 4.5 for total maternity ratio (this term, used by Nag, basically equals not only per-woman number of live births but also completed fertility, a prevailing demographic term), with 15% between 3.5 and 4.5 and 15% less than 3.5. The total fertility rate of the hunting-gathering !Kung San was calculated, in a thorough way, by Howell (1979) using two cohorts, one involving 62 women of completed reproductive age and one involving 166 women in their childbearing years; the rate for the former was 4.69 and that for the latter 4.25. These rates have been discussed in relation to nutrition, health status, and lactational practice (e.g. Howell, 1979; Konner and Worthman, 1980) as well as energetics (Bentley, 1985). All studies have focused on searching for reasons why the !Kung have maintained such a low fertility.

The population growth rate in the prehistoric period is still in debate. Reviewing archaeological and ethnographic information, however, Hassan (1973) suggests that the annual growth rate during the Neolithic was 0.001 on the average, with great fluctuation. In a global sense, it is appropriate to consider that the 0.002 growth rate of the Gidra is within the range of Hassan's estimate. In this connection, we cannot deny the possibility that ethnographic data tend to show a higher fertility or growth rate than the traditional level because of their treatment of living women only.

One of the factors that contributed to the low fertility of the Gidra was a sterility rate of more than 15%, which is higher than the average of 12% among Nag's (1962) samples. The reasons have not been fully clarified, although, despite Nag's emphasis on the high correlation between incidence of sterility and prevalence of venereal diseases, the Gidra's high sterility seems not to be attributable to venereal diseases; this will be discussed below in relation to the intra-population variation of DMR.

The bulk of the literature emphasizes that nonliterate human populations have practiced a variety of artificial methods of population control.

According to Nag's (1962) statistical analysis using "high" and "low" dichotomous categories, however, use of contraceptive devices and frequency of abortion are regarded as "low" in 42 of 47 societies and in 30 of 41 societies, and neither factor is associated with a "low" or "high" fertility level. Regarding infanticide, it is difficult to determine the extent of the practice in nonindustrial societies in general, because of the lack of cross-cultural analysis like Nag's on fertility controls. Although a high incidence of infanticide and its significant role in population regulation were reported, for example, among the Australian aborigines (Birdsell, 1968), Weiss (1972) suggests that infanticide is practiced on infants who are vulnerable owing to physical weakness or owing to the mother's difficulty in nursing; infanticide among the sago-gathering Sanio-Hiowe in Papua New Guinea is related to the latter (Townsend, 1971).

The role of artificial fertility and mortality controls in population regulation may have been overestimated. The population increase rate was low among the Gidra, who practiced no or little artificial control.

Finally, the 0.002 annual growth rate of the Gidra should be historically considered in relation to local conditions. Prehistorians suggest that the lowland Melanesians, in general, were characterized by a significant amount of human movement, even if the contemporary groups are geographically, culturally, and linguistically separate (Hope *et al.*, 1983). According to Swadling (1983: 31), the cultural identities of the big groups on the south coast of New Guinea might have emerged only in the last 300 years. If so, it is meaningful to estimate the origin of the population of the Gidra at 300 years B.P.

It is first necessary to estimate the population in 1945, after which the annual growth rate should have increased. The total number of living people (including out-migrants) in 1980 who were born in the Gidraland was 2230 (Chapter 20 in this volume). From a comparison of per-woman number of live births between groups A and B, the annual increase rate of the former was calculated at 0.0155; this rate is supposed not to have changed until 1980 (in the last few decades, fertility has increased and mortality has decreased, and the out-migration of young people has increased). From the above figures, the population in 1945, N_m, was calculated as follows:

$$r = \frac{1}{t} \cdot \ln\left(\frac{N_2}{N_1}\right),$$

where $r=0.0155$, $t=35$, $N_2=2230$, and $N_1=N_m$. As the result, N_m was 1296, or approximately 1300.

Since we have supposed that the annual growth rate of the Gidra prior to 1945 was constantly 0.002 and that their establishment dates back to

300 years B.P., the initial population (N_i) can be calculated using the same formula in which $r=0.002$, $t=300$, $N_2=1300$, and $N_1=N_i$. The result shows that N_i was 713, which seems to have the potential of maintaining population reproduction on the one hand (cf. Wobst, 1974) and establishing cultural identity on the other.

INTRA-POPULATION VARIATION

As is shown in Table 3, DMR varies when the women are broken down by their characteristics, i.e. birthplace, death or living place, experience of inter-village (or inter-population) migration, and number of marriages. First, DMR greatly differs between two groups categorized by number of marriages. A custom of the Gidra that a reproductive-aged woman observes only a short period of widowhood is suggestive of the weak effect of number of marriages *per se* on the different DMR. In my view, this difference reflects to a greater extent the different longevity of the women; in particular, women who survived only briefly after marriage might not have been few in number, and most of them would have belonged to the one-marriage category.

Second, the difference in experience of inter-village (or inter-population) migration is difficult to explain. One conceivable reason is differences in genetic properties. Although such indicators as an inbreeding coefficient have not been examined, attention should be paid to the fact that among the Gidra about 70% of marriage have been village-endogamous (Chapter 20 in this volume). On the other hand, of the 228 migrant women, 50 married men of different populations (linguistic groups), and the total number of their married daughters was 59; this ratio is equivalent to 1.18 DMR. This suggests that the inter-population migrants have a higher potentiality of reproduction than inter-village (intra-population) migrants.

Finally, the most striking difference in DMR was found in the breakdown by the women's birthplace and death or living place. This microenvironmental or microgeographical difference can be discussed from the viewpoint of health and disease. Concerning venereal diseases, which may have related to the high proportion of sterility, however, we have no evidence of their prevalence in the Gidra. If they had been a decisive factor in decrease in fertility, the coastal group should have been most affected. Also noted is that there has been no evidence of widespread use of antibiotics among the Gidra, at least before a decade ago.

In his review on demographically important diseases in Papua New Guinea, Riley (1983) emphasizes the effect of malaria, which not only kills, in particular, infants and young children but also depresses fertility. Although the Gidraland may have mesoendemicity owing to the low population density, there is a high possibility that malaria relates to the microgeographical difference in DMR since the northern as well as riverine villages abound in mosquitoes. Also emphasized by Riley is malnutrition, which increases susceptibility to infectious diseases. Our food consumption study in the four villages (one northern, one inland, one riverine, and one coastal) reveals that their intakes of energy and protein in 1981 were beyond the levels recommended by FAO/WHO (1973) but suggests marginal protein intake a few decades ago in the northern and riverine villages largely because of low consumption of land animals (Chapter 9 in this volume). On the other hand, our analysis of metal concentrations in hair reveals that zinc contents of married women in an inland village (Wonie) in 1971–1972 and particularly in a northern village (Rual) are deficient or marginal, and this might result in a high incidence of human central nervous system malformation (Chapter 16 in this volume; see also Ohtsuka and Suzuki, 1978).

The reduction of the inter-village difference from the breakdown by birthplace to that by death or living place implies that the above-mentioned factors exerted more influence in the pre-reproductive age (or before marriage) of the women than in their reproductive age. Although this aspect remains to be explored fully, it can at least be said that health and nutritional status in childhood or adolescence is significantly related to reproductive capability.

Whatever the major reasons are, the fertility or reproduction rate of the Gidra markedly differs microenvironmentally. In other words, the Gidra's territory of only 4000 km², the bulk of which is recognized as lowland plain, is heterogeneous in the sense of human reproduction or adaptation. This suggests the possibility that anthropological or prehistoric populations have had to survive by coping with minutely diverse habitats.

(*Ryutaro Ohtsuka*)

Chapter 19

Fertility and Mortality in Transition

Early (1985) questions whether reported forager fertility represents the traditional level and suggests a high possibility of underenumeration due to technically incomplete recording of births, recent change caused by prevalence of introduced diseases, and so on. However, there also exists a rather contradictory fact: that population increase rates reported from contemporary hunting-gathering and horticultural populations are, in general, markedly higher than the estimated rates for prehistoric populations, for instance, about 0.1% annual rate during the Neolithic (Hassan, 1973). Even the Dobe !Kung, whose fertility is recognized by many scholars including Early (1985) as extremely low, had a 0.263 annual increase rate (Howell, 1979).

To consider this problem in the Gidra, there are two basic conditions. First, as demonstrated in the previous chapter (Chapter 18), annual increase rate of the Gidra as a whole was about 0.2% in the past, approximately before 1950. Second, their population increase rate has been markedly higher than that level, judging from our observations. For example, the number of births which took place during almost one year, between our 1980 and 1981 survey periods, far exceeded that of deaths. This chapter aims to analyze the recent changes of fertility, mortality, and natural increase rates of the Gidra. This aspect is indispensable in understanding the present and the long-term adaptive mechanisms of the Gidra, and concurrently useful in considering demographic situations in anthropological populations like the Gidra, in relation to data collection and analysis.

As has been mentioned in Chapter 18, no reliable records of vital events are available for the Gidra, and even the people's exact age is still difficult to obtain. Thus, applicable methods to the changing fertility, mortality, and reproduction levels are very limited, and the expected results of the present analysis can compare only the differences between the fertility level of women born in approximately 1900 to 1920 and that of women born in approximately 1920 to 1940 and between the premarital mortality rates of the offspring born to the two groups of women.

However, this comparison is of great validity for the present purpose, since so-called modernization of the Gidra became prominent after WWII. Even in Daru, the capital of Western Province and the sole center of the Gidra people's modernization, marked development took place since WWII; for example, of the 28 spontaneous settlements which contained a large number of inhabitants in the 1970s and 1980s, only one had been established before WWII and most of the remainder in the late 1940s and 1950s (Hashimoto, 1983). Within the Gidraland, Christianity, primary school education, and a local government system, which can be recognized as major modernization items in the area, were introduced in the 1960s, even though the coastal and riverine villagers had occasional contact with Christian missions and other foreign influences in Daru in earlier times. Furthermore, out-migration to urban areas, which is a good measure of the degree of modernization in Papua New Guinea (e.g. Skeldon 1979; Ross, 1984), has occurred since the 1950s or 1960s in the coastal and riverine villages and since the 1970s in the inland and northern villages.

METHODS

The data, on which this chapter is based, were collected by two different ways. The first was genealogical records, used also in Chapter 18, which identified name, sex, clan, birthplace, living or death place, and marital, parent-child, and sibling relations (in biological terms) of all living members and their ancestors. This study was conducted in 1980, with additions and revisions made in 1981–82. The second data set (also used in Chapter 18) was collected in 1981–82 by interviews with ever-married women about their detailed childbearing histories and their offspring's vital events. This survey was carried out by four researchers including myself in four villages, i.e. Rual, Wonie, Ume, and Dorogori, in each of which one of us stayed for several months. All the subject women in the latter survey were involved in the former.

Married women were classified into groups, taking their generation into account, and two of these groups are analyzed in this chapter: the "younger" group consisting of women who completed reproduction and whose female offspring have not married, and the "older" group consisting of women whose female offspring have married but have not completed reproduction. (The "younger" group and the "older" group imply, respectively, group A and group B in Chapter 18.) Regarding judgment as to the completion of reproduction, the same criteria as in Chapter 18 were used; in short, when the last child had passed the first grade of the Gidra age-grade system, i.e. *sobijogbuga* for male and *sobi-*

jogngamugai for female, the mother was recognized to have completed reproduction. It is estimated that the birth years of "older" group ranged approximately from 1900 to 1920, and those of the "younger" group, from 1920 to 1940. Therefore, the bulk of their childbearings took place, respectively, from 1920 to 1950, and from 1940 to 1970.

The married women treated in this chapter numbered 266 for the "older" group and 236 for the "younger" group in the genealogical records, and 21 for "older" and 50 for "younger" in the interviewed records. In all analyses in this chapter, the subjects were not divided by the village locality because of the small number of subjects in the interview survey; however, all the 71 subjects had completed reproduction in our four intensively studied villages.

Because of the restrictions of our data for the Giara, as for nonliterate populations in general, the present analysis used a fertility measure called completed fertility, i.e. the cumulative fertility at the end of reproductive age (Pressat, 1985); the same measure was called total maternity ratio and was used by Nag (1962) for cross-cultural comparison. It is noted that the completed fertility values differ from those of total fertility rate, a common demographic term, which implies the sum of the age-specific fertility rates over reproductive ages for a particular period (usually one year), even though both are used for the same purpose.

Regarding the mortality rate, the present analysis treated premarital deaths among the offspring of the subject women. The interview records clearly determined the age-grade at which each individual died. Information collected in our repeated field surveys confirmed that the transfer from the first to the second age-grade in either sex took place at the age of eight (occasionally seven or nine). The estimated age of females at marriage, or entrance to the third grade (*kongajog*), had been fairly identical, at 17 to 19 years old, although since the 1970s this age has tended to lower according to, in particular, their schooling until the late teens. On the other hand, males' age at marriage greatly varied, although the mean marital age was estimated at 28 years from various information, for instance, the differences in marriage ages among male and female siblings. From the proportions of deaths before the specific ages among the offspring (by sex) of the interviewed women, the probabilities of death were compared to the levels on the General Pattern of the United Nations Model Life Table to estimate the mortality schedule according to age (United Nations, 1982).

FERTILITY

Table 1 shows the completed fertilities of women belonging to "older"

Table 1. Completed Fertility of Groups Based on Genealogical and Interview Records

Group (N)	Completed fertility
Genealogical	
Older: died (197)	3.010
Older: living (69)	3.362
Older: both (266)	3.102
Younger: died (53)	2.887
Younger: living (183)	4.546
Younger: both (236)	4.174
Interviewed	
Older (21)	3.714
Younger (50)	4.640

and "younger" groups, obtained from two different data sets; the subject women in the genealogical records are grouped into two according to whether they were alive or dead at the time of my investigation in 1980. Compared to the accuracy in the interviewed records, there was a high possibility that the genealogical records excluded offspring who had died in infancy or early childhood since those offspring were occasionally faded out of the villagers' memory, and this possibility was higher in the "older" group than in the "younger" group, and in the "died" sub-group than in the "living" sub-group. Such exclusions of offspring would result in underestimation of the fertility rate. At the same time, however, the difference of fertility figures between "died" and "living" sub-groups was largely attributable to the different longevity of the women in the two sub-groups. In particular, the great difference in completed fertilities between "living" and "died" sub-groups in the "younger" group in the genealogical records are largely due to this reason, since the childbearing histories of "younger" group should have still been well memorized by the villagers even if the women in question had deceased.

With some reservations about the accuracy of the figures in Table 1, the completed fertilities can be compared between the following pairs: "older: living" versus "younger: living," "older: both" versus "younger: both" in the genealogical records, and "younger" versus "older" in the interview records. The results are shown in Table 2, in which the difference in the completed fertilities was about 1.35 for the genealogical data and 1.25 for the interview data. It seems safely to recognize that the increase rate in the fertility between the two groups was 25% on the basis of only the interviewed records, or 30% from both records.

To seek the reasons for the increase of fertility, the mothers in each of the four groups, i.e. "older" and "younger" in the genealogical records

Table 2. Ratios of Completed Fertilities among Three Pairs

	Ratio
Genealogical	
[Younger: Living] / [Older: Living]	1.352
[Younger: Both] / [Older: Both]	1.346
Interviewed	
[Younger] / [Older]	1.250

and "older" and "younger" in the interview records, were broken down by the number of live births. The cumulative frequencies of the mothers in each group are illustrated in Fig. 1. Comparing the pairs of "older" and "younger" groups in either genealogical or interview records, a marked difference is found not in the proportions of mothers who had fewer than three live births, but in those of mothers who had several live births. Special note should be given to the fact that the proportion of sterile women changed to a lesser extent. It is natural to judge that the increase of childbearings of the mothers, who might have had several (particularly, three to five) children, in the older group's pattern, played the most significant role in fertility change as a whole, and this may have largely owed to improved nutritional and health status and prolonged longevity.

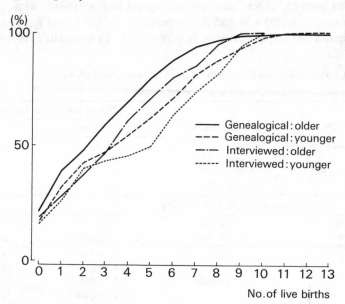

Fig. 1. Cumulative proportions of the women according to the number of live births. The same data are shown in Table 2 in Chapter 18.

MORTALITY

When we compared the proportion of premarital deaths among off-spring of the women in the interview survey with that among offspring of the women in the genealogical survey, the former was higher than the latter by about 10% in the offspring of "older" group, and less than 5% in the offspring of "younger" group. This discrepancy may have come from the omission of offspring, who died in infancy or early childhood, in the genealogical survey; in my judgments, in order to collect thorough information about the childbearings and the vital events of the offspring, indispensable was repeated interview by the researcher who obtained full rapport with the subject people, especially the women in question. Thus, the present analysis treats the proportion of premarital deaths among offspring of 71 women, the subjects of our interview survey.

The number of live births and number of premarital deaths are shown in the upper half of Table 3, which demonstrates that the premarital death rate has declined from "older" group to "younger" group. This becomes more prominent when the proportions of survivors at age 9 (l_9) for both sexes and either at age 19 (l_{19}) for females or at age 29 (l_{29}) for males are applied to the General Pattern of the United Nations Model Life Tables. As the result (the lower half of Table 3), l_9 and l_{19} of female children born to "older" mothers correspond to the patterns with e_0 (life expectancy at birth)$=38$ and 37, respectively, while l_9 and l_{29} of males correspond to the patterns with $e_0=39$ and 38. In contrast, l_9 of female

Table 3. Premarital Death of Children Born to Interviewed Women

Number of live births and number of premarital deaths

Group of			No. of premarital deaths	
mothers (N)	No. of live births		< 8 yrs	8+ yrs
Older (21)	Male	37	11	4
	Female	41	12	3
Younger (50)	Male	109	17	
	Female	121	28	

l_x and e_0 in the corresponding Model Life Table

Group of mothers	Sex of children	l_x (/100000)	e_0
Older	Male	$l_9 = 70818$	39
		$l_{29} = 59956$	38
	Female	$l_9 = 70426$	38
		$l_{19} = 64440$	37
Younger	Male	$l_9 = 84340$	53
	Female	$l_9 = 76955$	45

and male children of "younger" group mothers accords, respectively, with the patterns with $e_0 = 45$ and 53. Although the reasons why the difference of e_0 between the male and female children in the latter group increased have remained to be explained, the premarital mortality has markedly declined from the "older" group to the "younger" group, with life expectancy at birth increasing by approximately ten years.

If the age-specific mortality schedule of the female offspring of the "younger" group follows the Model Life Table with $e_0 = 45$, the corresponding l_{19} is 74638/100000. This figure is higher than that for the female offspring of the "older" group (64440/100000) by about 16%; in other words, the proportion of daughters who marry will increase by this proportion.

POPULATION INCREASE

As mentioned in Chapter 18 of this volume, annual increase rate of a population can theoretically be calculated from daughter-mother ratio (DMR), per-mother number of daughters who were born to a group of mothers during their lifetimes and who married, with some methodological reservations discussed in Chapter 18. The DMR of the "older" group (group B in Chapter 18) was 1.060, or 1.045 if the average value of the "older" group and the two preceding groups (groups C and D in Chapter 18) combined was applied. These DMR figures corresponded to annual increase rates of 0.26 and 0.19%. According to the previous calculations on the differences between the "older" and "younger" groups, fertility increased by 25–30% and the proportion of daughters who marry would increase by 16%. Thus, the DMR of the "younger" group can be estimated by 1.045 (or 1.060) × 1.25 (or 1.30) × 1.16. The results have proved 1.5–1.6 of DMR. These DMR figures can be converted to annual increase rate at 1.8–2.1%, or around 2%.

The change in the estimated annual increase rate by about ten times, from 0.2–0.25% to 2%, was observed in the inter-generational replacement between the "older" and the "younger" groups, and thus should have taken place in the last several decades. There is no doubt that the increase of fertility and decrease of mortality progressed along with the modernization process, even though the factors remain to be fully distinguished. In my speculative view, the most important for the decrease of mortality was the spread of medical service, which reduced infant deaths by means of practice of vaccination and adult and child deaths from malaria by providing anti-malarial tablets. On the other hand, the increase of fertility may have largely owed to the betterment of nutritional status through technological improvements in subsistence activities and gradual involvement in the cash economy (e.g. Chapter 9 in this volume). Whether the naturally increased population of the Gidra has met with the limit of carrying capacity of their land is still questioned, although, as mentioned in Chapter 20, the modernization has also triggered out-migration of the people to urban areas and consequently the number of inhabitants in their land has been fairly stable in the last one or two decades.

Whatever the reasons are, he demographic situation, which influences the people's adaptive system on the one hand and, on the other, is influenced by it, has drastically changed in the last several decades in the Gidra. This is important in understanding even the current adaptive mechanisms, since the people, particularly the aged, grew up in a situation which differed from the present one, and their sociocultural and behavioral norms may have been linked with the old situation. At the same time, it can also be said that the Gidra, like many other anthropological populations, have changed and will continuously change their way of life in the course of modernization.

(Ryutaro Ohtsuka)

Chapter 20

Inter- and Intra-Population Migration

Like an animal population, a human population can be defined genetically as "a reproductive community of sexual and cross-fertilizing individuals which share in a common gene pool" (Dobzhansky, 1968), or ecologically as "a collective group of organisms (of the same species) occupying a particular space" (Odum, 1971: 162). When human populations conform to both definitions they are the ideal unit of human ecology study, because such units form the basis of adaptation and survival not only in short duration (e.g. one-year duration) but also through generations (Suzuki, 1980; Ohtsuka, 1983). This viewpoint has been prevalent in human genetic studies (e.g. Harpending and Jenkins, 1974; Neel and Weiss, 1975; Skolnick et al., 1976). This chapter treats inter- and intra-population migrations of the contemporary Gidra people, including those who were born in the Gidraland and have migrated out of it and those who were born outside the Gidraland and have migrated to it, with special reference to the structure of the Gidra population.

SUBJECTS AND METHODS

During our two field investigations in 1980 and in 1981–82, genealogical-demographic data were collected (see Chapter 18 in this volume). In this study, the birth and living places were identified by the name of village or town. In processing the data, however, those outside the Gidraland were classified into larger units: (1) the villages of seven adjacent linguistic groups (Wurm, 1971); (2) Daru, and (3) "other" for places beyond the adjacent lands but within the Western Province; and (4) Port Moresby (the national capital), and (5) "other" for the localities beyond Western Province. To classify people by age, this paper follows the traditional age-grade system, which classifies the males into six grades and the females into four grades, although due to the small number, the oldest two groups for males, i.e. *nanyuruga* and *miid*, are treated together as *nanyuruga* (see frontispiece).

Here the word, migration, is defined as follows: any individual whose

219

birth and living places differ is classified as having migrated from the former to the latter, while the others are classified as not having migrated. Except for recent out-migration to urban areas, the villagers usually grew up in their birthplaces, and at the time of marriage some of them migrated from their birthplaces and others stayed in them.

DEMOGRAPHIC OVERVIEW

Three summary tables (Tables 1, 2, and 3) demonstrate the basic demographic characteristics of the subjects. As shown in Table 1, the *de facto* population of the Gidraland in 1980 numbers 1850, of which 1791 (96.8 %) were born within it. The very small proportion of in-migrants is of special importance for assessing the Gidra as a population, and this will later be discussed.

Table 2 gives the breakdown of the people who were born in the Gidraland by living place. In contrast with the 59 in-migrants to the Gidraland, out-migrants number 439. Most destinations of the out-migrants are urban areas beyond the territories of the adjacent linguistic groups. Classified by sex and age-grade, the proportion of in-migrants is higher among married persons, especially the females, while the proportion of out-migrants is highest among the *kewalbuga*, second highest among the *rugajog*, and third highest among the *kongajog*. A very few cases of out-

Table 1. The Present Inhabitants of the Gidraland by Age-Grade[a] and by Birthplace

	Male					Female				
Birthplace	NR	RJ	KB	YB	SB	NK	KJ	NB	SN	Total
Gidraland[b]										
Same village	55	213	150	140	221	67	206	219	266	1537
Other villages	23	39	11	5	14	33	119	5	5	254
Adjacent lands[c]	1	8	0	0	1	7	15	1	0	33
Within Western Province										
Daru	0	0	0	0	2	0	0	2	1	5
Other places	0	2	0	2	1	1	4	0	1	11
Beyond Western Province										
Port Moresby	0	0	0	0	0	0	0	1	0	1
Other places	0	4	0	1	2	0	1	0	1	9
Total	79	266	161	148	241	108	345	228	274	1850

[a] Abbreviations of age-grades; NR: *nanyuruga*, RJ: *rugajog*, KB: *kewalbuga*, YB: *yambuga*, SB: *sobijogbuga*, NK: *nanyukonga*, KJ: *kongajog*, NB: *ngamugaibuga*, SN: *sobijobngamugai*. *Miid* are included in *nanyuruga*.

[b] The Gidraland is categorized into two. "Same village" applies to a person who was born and lives in the identical village.

[c] "Adjacent lands" imply the lands of seven linguistic groups around the Gidraland: Bine, Gizra, Agob, Idi, Paswam, Lewada-Dewara, and Tirio.

Table 2. The Living People Who Were Born in the Gidraland, Classified by Age-Grade and by Living Place[a]

Living places	Male					Female				Total
	NR	RJ	KB	YB	SB	NK	KJ	NB	SN	
Gidraland										
Same village	55	213	150	140	221	67	206	219	266	1537
Other villages	23	39	11	5	14	33	119	5	5	254
Adjacent lands	3	5	3	4	6	7	18	3	4	53
Within Western Province										
Daru	1	20	26	2	2	2	26	15	0	94
Other places	0	27	9	0	5	3	23	2	10	79
Beyond Western Province										
Port Moresby	1	39	82	0	0	0	21	6	0	149
Other places	0	16	33	0	0	0	9	4	2	64
Total	83	359	314	151	248	112	422	254	287	2230

[a] Abbreviation of age-grades and categorization of place names are as mentioned in footnotes to Table 1.

migration of *nanyuruga* and *nanyukonga* accord with the information obtained in our field work that most out-migration to urban areas has occurred since the 1950s or 1960s in the coastal and riverine villages and since the 1970s in the inland villages.

The breakdown of inter-village migration proves two observations (Table 3). First, of the 78 two-village pairs, more than half (58%) have no migration. Second, the balance between in-migrants and out-migrants differs from village to village; this balance and its ratio to the number of people who were born and remain in the Gidraland are shown in Table 4. This ratio (called rate of inter-village migration balance) is positive for the coastal and riverine villages and is negative for most inland villages. An important question is whether the differing rate has changed the population distribution among the 13 villages.

The above overview has highlighted two major discussion points. The first is to re-examine whether the Gidra group has the characteristics of a population. The associated and more important problem is the migratory relationship of the 13 villages as a major component of the demographic infrastructure of the Gidra. The second point is the recently increasing out-migration to urban areas in the light of the population dynamics of the Gidra as a whole.

THE GIDRA AS A POPULATION

In genetic evaluation of a population, emphasis is placed on intermarriage with individuals from other populations, although there is no consensus on the intermarriage rate up to which a human aggregation can be

Table 3. The People Who Were Born and Have Remained in the Gidraland, Classified by Birth and Living Villages

Birth village	Living village													Total
	RU	KA	IA	WI	PO	GA	WO	KU	UM	WU	WG	AB	DO	
Rual	95	3	0	0	0	0	0	0	0	0	0	0	0	98
Kapal	5	108	9	1	1	0	0	0	0	0	0	0	0	124
Iamega	0	6	224	10	0	0	16	0	1	8	0	0	0	265
Wipim	0	0	4	121	7	4	1	0	0	0	0	0	0	137
Podare	0	0	1	19	106	5	3	0	1	0	0	0	0	135
Gamaeve	0	0	1	1	10	136	1	0	7	0	0	0	0	156
Wonie	0	0	0	2	6	3	84	2	3	4	0	0	0	104
Kuru	0	0	0	0	0	1	3	120	14	2	11	17	8	176
Ume	0	0	0	0	0	2	0	3	142	0	0	0	3	150
Wuroi	0	0	1	0	0	0	0	0	0	83	12	2	0	98
Woigi	0	0	0	0	0	0	0	0	0	7	108	4	5	124
Abam	0	0	0	0	0	0	0	4	0	0	1	114	5	124
Dorogori	0	0	0	0	0	0	0	1	2	0	0	1	96	100
Total	100	117	240	154	130	151	108	130	170	104	132	138	117	1791

Table 4. Balance of Inter-Village Migrants by Village and Its Ratio to the Residents Who Were Born and Remain in the Gidraland

Village	No. of births	Balance of inter-village migrants	Rate
Inland[a]			
Rual	98	2	0.020
Kapal	124	−7	−0.056
Iamega	265	−25	−0.094
Wipim	137	17	0.124
Podare	135	−5	−0.037
Gamaeve	156	−5	−0.032
Wonie	104	4	0.038
Kuru	176	−46	−0.261
Riverine			
Ume	150	20	0.133
Wuroi	98	6	0.061
Woigi	124	8	0.065
Abam	124	14	0.113
Coastal			
Dorogori	100	17	0.170

[a] *Inland* includes Rual and Kapal villages, which are categorized in northern group in other chapters of this volume.

recognized as a population. Among the Gidra, 43 of the 798 married members were born outside their territory (Table 1), that is, a 0.054 migration rate. This rate roughly equals 0.05, which Bodmer and Cavalli-Sforza (1968) gave as the maximum for linear stabilizing selection in a theoretical model. They pointed out, however, that the migration rate in most real populations would be higher than 0.05. The migration rate of the Gidra's 1850 individuals can be judged to satisfy the nature of a population. In this connection, the number of people involved in the examined unit is relevant (Bodmer and Cavalli-Sforza, 1974); for example, Brainard (1982) reported a very low migration rate (0.011) among the married members of the nomadic pastoral Turkana, but because the population numbers about 200,000 individuals in total, such a low rate may have occurred naturally.

Tables 1 and 2 demonstrate higher inter-village migration rate among the married, especially the females. This reflects the custom of residential moving of spouses at marriage. In Gidra society virilocal residence is popular, while uxorilocal residence also occurs. Figure 1 exhibits the relationship between the geographical distance of each two-village pair and the inter-village migration rate of the married, which is defined, for example between villages A and B, as follows:

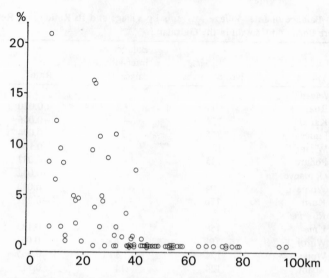

Fig. 1. The relation between inter-village migration rate of the married (on the ordinate) and the geographical distance (on the abscissa).

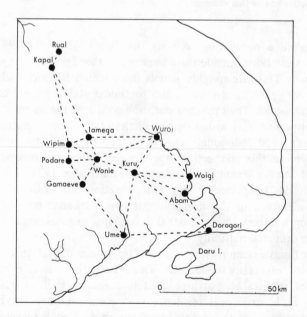

Fig. 2. The location of 13 Gidra villages. The dotted lines show the path (including canoe-travel) network.

$$\frac{1}{2} \times \left[\frac{\text{Number of the married of A, who were born in B}}{\text{Total number of the married of A}} \right.$$

$$\left. + \frac{\text{Number of the married of B, who were born in A}}{\text{Total number of the married of B}} \right].$$

Figure 1 discloses that inter-village migration seldom occurs between villages more than 40 km apart, although the rates of the remaining (less than 40 km) two-village pairs do not clearly correlate with the distance. To reassess the inter-village relation, each two-village pair is labeled by the scored contiguity of the (foot or canoe) path network in the Gidraland (Fig. 2). A contiguity score 1 implies that the two villages are linked by a path with no village in between; a score 2 or more is the number of the in-between villages plus 1. The means and SDs of the migration rates of the village-pairs by the contiguity score are shown in Table 5, which illustrates that the larger the score between 1 and 3, the lower—by more than ten times per point—the migration rate. Among the village-pairs with contiguity scores 4 and 5, no migration occurs.

It is difficult to organize all marriages within a single village of 100 to 250 dwellers. The means and SDs of intra-village marriage rates among the 13 villages are 0.7068 ± 0.1254. The rates are slightly higher than those observed among nine sub-populations of the hunting-gathering !Kung who were recognized to lack subdivision and inbreeding (Harpending and Jenkins, 1974). Theoretically, the average number of inhabitants of a single Gidra village, when all of them find mates, can be calculated as follows: $1/0.7068 \times 1850/13$, that is, about 201. (If we take the 5.4% of in-migrants from other populations into account, this figure increases to 213.) This size is in the lower part of the range between 175 and 475 which was calculated from Monte Carlo simulation as the minimum equilibrium size for hunter-gatherers in order to provide all members with mates upon reaching maturity (Wobst, 1974).

The rest (about 30%) of the mates are provided mainly from other Gidra villages, and Table 5 shows that most of them are inter-village

Table 5. Inter-Village Migration Rates of the Married by the Contiguity Score of-Two-Village Pairs

Contiguity Score	N	Mean ± SD
1	23	0.0762 ± 0.0528
2	28	0.0056 ± 0.0115
3	17	0.0004 ± 0.0017
4	7	0
5	3	0

Table 6. Number of Out-Migrants to Urban Areas and Its Rate to Births by Village

Village	No. of births	No. of out-migrants to urban areas	Rate
Inland			
Rual	111	9	0.081
Kapal	155	22	0.142
Iamega	306	40	0.131
Wipim	159	21	0.132
Podare	182	30	0.165
Gamaeve	205	43	0.210
Wonie	117	13	0.111
Kuru	187	11	0.059
Riverine			
Ume	193	32	0.166
Wuroi	139	41	0.295
Woigi	156	31	0.200
Abam	156	31	0.200
Coastal			
Dorogori	164	62	0.378

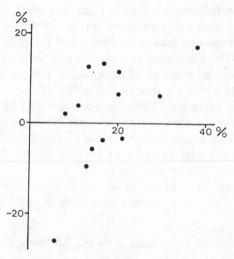

Fig. 3. The relation between rate of inter-village migration balance (on the ordinate) and out-migration rate to urban areas (on the abscissa) for the 13 villages.

migrants within the village-pairs with contiguity scores of 1. Another feature of the inter-village migration is the lack of subdivision of villages; this is typically evidenced by the fact that there is no village-pair with a contiguity score of 1 in which no inter-village migration occurs.

In summary, the Gidra as a population consist of villages or sub-populations. The Gidra population is maintained by an overlapping inter-marriage network of adjacent villages throughout the territory.

OUT-MIGRATION TO URBAN AREAS

A current characteristic of Gidra demography is a large number of out-migrants to urban areas. The number differs from village to village. Table 6 shows number of each village's out-migrants and its rate to total births; this rate is called the out-migration rate (M_O). The M_O is higher in the coastal and riverine villages than in the inland villages. One effective way to understand out-migration in the context of the population dynamics of the Gidra is to relate M_O to the rate of inter-village migration balance (M_B), as shown in Table 4.

The scatter diagram of M_O and M_B for the 13 villages is shown in Fig. 3. When we calculate the correlation coefficient between the two variables, a simple correlation coefficient (r) is 0.589 ($p < 0.05$); applying rank correlation coefficients, Spearman's $r_S = 0.565$ ($p < 0.05$) and Kendall's $r_K = 0.448$ ($p < 0.05$). This significant correlation implies that the higher the out-migration rate to urban areas, the higher the rate of inter-village migration balance. Thus, it is reasonable to judge that the large number of out-migrants from the more modernized coastal and riverine villages tend to be offset by inter-village migrants from the less modernized inland villages.

(Ryutaro Ohtsuka)

Appendix : Composition of the Gidra Foods (per 100 g Edible Portion)

Common name	Scientific name	Mois-ture (g)	Energy (kcal)	Protein (g)	Fat (g)	Carbo-hydrate (g)	Fiber (g)	Ash (g)	Na (mg)
Sagos and tubers									
sago flour	Metroxylon sp.	42.0	230	0.5	0.0	56.8	0.3	0.3	13.1
sago flour	Metroxylon sp.	46.9	211	0.3	0.0	52.3	0.1	0.1	8.1
areca flour	Areca sp.	53.9	197	2.7	0.5	44.5	1.0	3.3	12.6
taro	Colocasia esculenta	62.8	146	1.8	0.1	33.8	0.7	0.8	0.9
taro	Colocasia esculenta	68.8	121	1.9	0.1	27.5	0.7	1.1	0.4
taro	Colocasia esculenta	68.1	124	1.7	0.1	28.3	0.7	1.1	0.7
elephant-foot yam	Amorphophallus campanulatus	76.7	89	1.5	0.1	19.1	1.4	1.2	2.3
yam	Dioscorea alata	75.5	95	1.4	0.1	21.3	0.7	1.1	6.2
sweet potato	Ipomoea batatas	80.6	77	1.1	0.7	15.5	1.1	1.1	19.2
cassava[a]	Manihot esculenta	64.4	140	0.3	0.1	33.6	0.8	1.0	4.0
cassava[a]	Manihot esculenta	60.4	157	0.4	0.2	37.4	1.0	0.9	1.8
cassava[a]	Manihot esculenta	67.2	—	—	—	—	—	—	4.9
cassava[a]	Manihot esculenta	49.7	—	—	—	—	—	—	10.5
pumpkin[a, b]	Cucurbita sp.	86.2	53	1.0	0.1	11.2	0.8	0.7	1.1
Fruits and nuts									
banana	Musa sp.	78.1	84	0.8	0.1	19.6	0.3	1.1	0.9
banana	Musa sp.	76.8	88	0.9	0.0	20.7	0.3	1.3	0.9
banana	Musa sp.	74.8	113	0.9	0.9	23.3	2.1	0.7	0.6
banana	Musa sp.	68.3	136	1.1	1.6	28.9	0.4	1.1	0.7
banana	Musa sp.	74.7	103	1.2	0.7	22.6	0.4	1.0	1.0
banana	Musa sp.	78.2	88	0.5	0.1	21.2	0.2	0.6	0.5
banana	Musa sp.	64.3	173	1.5	2.9	32.8	2.5	0.9	0.8
banana	Musa sp.	66.9	152	1.1	1.5	30.7	2.7	0.9	0.6
banana	Musa sp.	78.2	101	1.0	1.1	19.5	2.2	0.9	1.2
papaya[a]	Carica papaya	96.1	—	—	—	—	—	—	4.7
papaya[a]	Carica papaya	82.5	—	—	—	—	—	—	20.3
pineapple[a]	Ananas comosus	91.7	—	—	—	—	—	—	0.8
pineapple[a]	Ananas comosus	92.3	—	—	—	—	—	—	1.1
mango	Mangifera indica	86.2	55	0.4	0.2	11.5	1.4	0.3	2.2
cycad	Cycas circinalis	54.0	201	5.9	2.1	39.0	0.5	4.3	1.9
lotus	Nelumbo nucifera	72.9	106	6.9	0.9	16.2	1.3	1.7	1.4
galip	Canarium vitiense	59.2	226	2.0	14.5	12.4	9.6	2.4	54.0
jointfir	Gnetum gnemon	73.0	107	2.9	0.4	19.8	3.1	0.8	0.9
coconut : dry flesh[a]	Cocos nucifera	64.5	—	—	—	—	—	—	7.1
dry flesh[a]	Cocos nucifera	54.6	337	3.6	28.5	7.8	8.7	1.0	10.5
dry flesh[a]	Cocos nucifera	43.3	—	—	—	—	—	—	16.8
dry flesh[a]	Cocos nucifera	48.6	348	3.6	29.4	2.2	15.0	1.2	13.4
immature flesh[a]	Cocos nucifera	—	—	—	—	—	—	—	37.8
dry liquid[a]	Cocos nucifera	—	—	—	—	—	—	—	22.3
dry liquid[a]	Cocos nucifera	—	—	—	—	—	—	—	32.5
dry liquid[a]	Cocos nucifera	—	—	—	—	—	—	—	48.3
dry liquid[a]	Cocos nucifera	—	—	—	—	—	—	—	57.3
immature liquid[a]	Cocos nucifera	—	—	—	—	—	—	—	12.4
Leaves									
abika[c]	Abelmoschus manihot	86.2	51	3.5	0.7	5.5	2.3	2.0	31.1
jointfir	Gnetum gnemon	83.1	65	4.7	0.5	5.5	4.8	1.4	12.2
bedum[d]	Ormocarpum orientale	75.8	93	6.5	1.0	11.8	2.8	2.2	25.6
Purchased plant foods									
rice	Oryza sativa	14.1	343	5.6	0.5	78.9	0.2	0.7	2.1
wheat flour	Triticum sp.	12.9	342	10.3	1.1	72.6	0.2	2.9	680.6
biscuit	—	8.8	384	6.8	5.2	77.5	0.3	1.4	354.0
biscuit	—	10.8	359	12.5	1.6	73.6	0.3	2.0	566.2
Land mammals									
pig	Sus scrofa	77.0	90	20.5	0.7	0.5	—	1.3	53.1
pig	Sus scrofa	76.5	92	19.6	0.7	1.8	—	1.5	54.9
pig (liver)	Sus scrofa	68.5	138	22.0	3.8	3.9	—	1.8	108.0
deer	Cervus timorensis	75.6	106	19.3	2.8	1.0	—	1.3	41.0
deer	Cervus timorensis	77.2	88	18.3	0.4	2.9	—	1.1	32.5
deer (liver)	Cervus timorensis	76.8	100	14.9	2.5	4.4	—	1.3	68.3
wallaby	Walabia agilis	78.3	85	19.8	0.5	0.3	—	1.2	52.5
wallaby	Thylogale sp.	75.1	99	23.2	0.7	0.0	—	1.1	86.3
wallaby (liver)	Thylogale sp.	81.9	73	15.2	0.9	0.9	—	1.0	72.2
wallaby	Dorcopsis veterum	75.7	97	21.5	1.0	0.5	—	1.3	62.1
bandicoot	Echimipera ?rufescens	76.0	96	21.4	1.1	0.0	—	1.3	82.6
Birds									
cassowary	Casuarius casuarius	74.8	109	21.1	2.4	0.7	—	1.1	65.1
cassowary (fat layer)	Casuarius casuarius	38.3	467	16.9	44.4	0.0	—	0.4	178.5
duck	Dendrocygna guttata	76.3	101	17.8	2.1	2.6	—	1.2	73.3
brush turkey	?	74.5	104	20.5	1.3	2.5	—	1.3	57.2
pigeon	?	73.0	118	20.0	3.1	2.6	—	1.4	86.8
Reptiles									
crocodile	Crocodilus porosus	80.2	82	15.9	1.3	1.6	—	1.0	47.4
monitor lizard	Varanus sp.	73.0	101	23.8	0.6	0.0	—	2.5	53.8
monitor lizard	Varanus sp.	—	—	—	—	—	—	—	39.6

228

Mg (mg)	Al (mg)	P (mg)	K (mg)	Ca (mg)	V (μg)	Cr (μg)	Mn (mg)	Fe (mg)	Ni (μg)	Cu (mg)	Zn (mg)	Sr (mg)	Cd (μg)	Hg (μg)	Pb (μg)
9.1	2.20	8.9	60.3	10.2	n.d.	7.3	1.847	5.94	n.d.	0.040	0.11	0.100	n.d.	0.1	n.d.
9.6	1.66	9.5	61.8	9.6	n.d.	12.9	2.840	3.04	n.d.	0.120	0.11	0.054	n.d.	0.3	n.d.
17.2	1.53	7.9	38.3	10.7	n.d.	24.1	0.361	7.05	n.d.	0.016	0.92	0.913	n.d.	0.1	n.d.
44.7	n.d.	64.5	348.4	19.8	n.d.	10.2	0.787	2.15	n.d.	0.262	1.71	0.377	n.d.	1.0	n.d.
42.0	n.d.	46.5	345.5	14.0	n.d.	6.4	0.370	0.81	n.d.	0.212	0.56	0.268	n.d.	0.1	n.d.
47.4	n.d.	39.0	349.5	16.1	n.d.	13.3	0.247	1.51	n.d.	0.258	1.38	0.306	n.d.	—	1.69
36.9	0.53	59.7	562.1	13.3	n.d.	17.7	0.241	1.16	n.d.	0.130	1.62	0.053	n.d.	0.1	n.d.
17.8	0.15	58.0	451.2	14.2	n.d.	4.2	0.019	0.44	n.d.	0.147	0.49	0.028	n.d.	0.2	n.d.
20.4	0.73	48.6	358.6	20.9	n.d.	21.2	0.251	0.62	n.d.	0.151	0.21	0.305	n.d.	0.5	n.d.
22.1	1.18	40.8	525.2	15.0	n.d.	n.d.	0.109	0.93	n.d.	0.066	0.29	0.279	n.d.	—	n.d.
40.7	0.71	38.5	504.6	22.1	n.d.	n.d.	0.076	0.75	n.d.	0.060	0.56	0.202	n.d.	—	n.d.
31.7	0.40	80.7	708.3	19.9	n.d.	n.d.	0.108	0.47	n.d.	0.075	0.79	0.162	n.d.	—	n.d.
43.5	0.13	35.4	282.4	39.8	n.d.	n.d.	0.161	0.25	n.d.	0.058	0.43	0.372	n.d.	—	n.d.
12.4	1.15	22.0	382.5	8.1	n.d.	n.d.	0.105	1.03	n.d.	0.050	0.19	0.137	n.d.	—	n.d.
44.8	0.96	43.7	437.3	13.6	n.d.	9.2	2.375	2.93	n.d.	0.052	0.26	0.040	n.d.	0.1	n.d.
33.4	2.31	23.5	331.7	4.3	n.d.	11.6	0.089	1.74	n.d.	0.078	0.27	0.078	n.d.	0.1	n.d.
25.3	0.53	19.3	282.0	2.9	n.d.	46.5	0.130	2.14	n.d.	0.045	0.10	0.027	n.d.	0.2	n.d.
37.5	0.80	33.3	396.1	3.0	n.d.	21.4	0.183	3.09	n.d.	0.074	0.27	0.027	n.d.	—	n.d.
33.6	1.01	27.0	454.7	5.1	n.d.	18.3	0.156	3.71	n.d.	0.068	0.39	0.027	n.d.	—	3.49
39.4	0.96	23.4	286.8	3.4	n.d.	24.5	0.238	1.24	n.d.	0.023	0.14	0.049	n.d.	—	n.d.
48.9	0.51	28.0	336.7	2.5	n.d.	7.5	0.449	19.96	n.d.	0.046	0.22	0.033	n.d.	—	n.d.
39.5	n.d.	33.1	337.6	2.1	n.d.	3.1	0.210	1.17	n.d.	0.080	0.18	0.042	n.d.	n.d.	n.d.
33.6	0.62	28.2	354.4	6.4	n.d.	9.7	0.166	1.15	n.d.	0.036	0.21	0.039	n.d.	—	2.24
29.1	2.75	19.9	129.4	31.4	n.d.	n.d.	0.133	1.95	n.d.	0.031	0.12	0.326	n.d.	—	n.d.
10.7	0.39	8.5	179.2	10.8	n.d.	n.d.	0.044	0.54	n.d.	0.087	0.10	0.066	n.d.	—	n.d.
27.5	0.90	7.1	72.5	19.8	n.d.	n.d.	0.246	1.23	n.d.	0.061	0.14	0.102	n.d.	—	n.d.
66.4	0.46	10.3	138.9	83.3	n.d.	n.d.	2.598	0.44	n.d.	0.050	0.14	0.155	n.d.	—	n.d.
17.5	1.03	13.7	114.3	25.6	n.d.	12.6	0.412	2.14	n.d.	0.072	0.23	0.087	n.d.	0.4	2.13
35.9	0.36	74.6	137.0	13.1	n.d.	17.0	1.527	36.28	74	0.266	0.53	0.069	n.d.	1.4	n.d.
83.9	0.21	187.0	366.8	98.6	n.d.	9.0	5.887	9.89	203	0.501	1.36	1.034	n.d.	0.7	n.d.
79.2	n.d.	20.6	605.8	240.8	n.d.	23.5	22.735	13.61	n.d.	0.217	0.18	1.395	n.d.	2.6	n.d.
30.4	0.21	44.0	275.3	23.0	n.d.	26.5	1.336	0.77	n.d.	0.115	0.35	0.128	n.d.	0.8	n.d.
44.7	1.19	73.5	451.0	6.9	n.d.	n.d.	1.245	2.14	n.d.	0.324	0.60	0.077	n.d.	—	n.d.
54.2	0.59	92.7	421.4	12.8	n.d.	n.d.	0.870	2.81	n.d.	0.405	0.88	0.053	n.d.	—	n.d.
64.0	0.50	155.7	464.9	6.9	n.d.	n.d.	0.902	1.80	n.d.	0.528	0.66	0.009	n.d.	—	n.d.
53.1	0.29	89.2	503.5	12.6	n.d.	n.d.	0.801	1.44	n.d.	0.554	0.72	0.028	n.d.	—	n.d.
41.1	n.d.	45.8	205.3	10.3	n.d.	n.d.	1.118	0.62	n.d.	0.059	0.20	0.043	n.d.	—	n.d.
8.3	n.d.	5.1	183.3	30.5	n.d.	n.d.	0.798	0.05	n.d.	n.d.	0.03	0.140	n.d.	—	n.d.
9.9	n.d.	10.6	228.5	32.6	n.d.	n.d.	0.249	0.02	n.d.	n.d.	0.05	0.055	n.d.	—	n.d.
11.3	n.d.	14.0	258.7	12.8	n.d.	n.d.	0.153	n.d.	n.d.	0.009	0.01	0.002	n.d.	—	n.d.
15.0	n.d.	6.2	224.1	40.8	n.d.	n.d.	0.339	0.03	n.d.	0.014	0.05	0.034	n.d.	—	n.d.
10.8	n.d.	7.8	135.6	16.3	n.d.	n.d.	0.839	0.01	n.d.	0.005	0.05	0.013	n.d.	—	n.d.
174.5	0.39	36.9	320.1	305.4	71	30.3	2.660	4.86	52	0.111	0.92	1.677	n.d.	0.4	1.57
47.7	0.46	101.0	478.4	85.6	n.d.	7.0	3.256	1.40	n.d.	0.103	0.48	0.262	n.d.	0.6	n.d.
149.2	1.17	64.5	285.5	430.8	40	4.4	8.342	10.38	193	0.267	0.60	3.124	n.d.	0.8	n.d.
29.1	n.d.	96.0	97.5	5.1	n.d.	n.d.	1.546	0.20	n.d.	0.147	0.96	0.008	n.d.	—	n.d.
31.1	n.d.	593.4	136.3	18.4	n.d.	24.3	0.979	0.97	n.d.	0.118	0.77	0.167	n.d.	—	n.d.
19.7	n.d.	75.2	133.2	36.4	n.d.	17.4	0.756	0.60	n.d.	0.076	0.51	0.136	n.d.	—	n.d.
36.3	n.d.	174.8	148.5	18.0	n.d.	18.1	0.928	1.48	n.d.	0.146	0.73	0.189	n.d.	—	n.d.
25.8	2.49	257.3	416.2	3.8	n.d.	14.3	n.d.	14.62	n.d.	0.087	6.36	0.008	n.d.	2.9	n.d.
26.9	2.76	253.3	420.2	4.2	n.d.	12.9	0.026	4.29	n.d.	0.088	4.46	0.010	n.d.	—	n.d.
19.8	1.47	411.0	294.1	3.9	n.d.	34.8	0.320	111.03	n.d.	0.675	5.28	0.007	n.d.	—	32.35
29.2	0.99	248.2	436.1	3.7	n.d.	30.8	0.017	4.67	n.d.	0.181	3.39	0.011	n.d.	1.0	n.d.
28.2	0.15	236.1	391.3	3.6	n.d.	7.1	0.012	2.32	n.d.	0.113	1.91	0.005	n.d.	0.5	n.d.
16.4	0.67	302.4	292.7	5.1	n.d.	25.9	0.232	15.88	n.d.	22.891	3.03	0.016	n.d.	3.0	4.67
28.4	0.72	251.2	357.9	3.8	n.d.	9.5	0.030	4.82	n.d.	0.128	1.93	0.005	n.d.	0.4	n.d.
21.6	1.20	200.1	295.3	5.5	n.d.	23.4	0.062	13.66	n.d.	0.127	2.98	0.009	n.d.	0.8	n.d.
13.7	0.32	252.5	222.0	3.6	n.d.	5.9	0.331	8.76	n.d.	0.291	3.08	0.005	n.d.	—	n.d.
28.0	1.12	249.5	352.6	4.9	n.d.	15.0	0.022	2.88	n.d.	0.101	2.24	0.011	n.d.	1.1	n.d.
31.2	1.83	279.9	467.5	6.5	n.d.	9.9	0.088	3.12	n.d.	0.147	2.07	0.019	n.d.	1.1	n.d.
29.2	0.77	259.4	368.9	4.4	n.d.	60.1	0.037	5.72	n.d.	0.153	4.19	0.006	n.d.	0.7	1.02
25.0	2.21	369.9	446.2	8.4	n.d.	55.2	n.d.	5.59	n.d.	0.039	2.43	0.006	n.d.	—	8.79
30.4	1.43	251.9	366.9	12.2	n.d.	19.6	0.121	9.54	n.d.	0.173	2.03	0.039	n.d.	0.1	104.08
33.8	0.48	295.5	437.4	5.1	n.d.	32.6	0.032	4.27	n.d.	0.137	1.14	0.015	n.d.	7.7	14.01
32.2	0.75	293.3	360.5	5.9	n.d.	14.7	0.074	5.58	n.d.	0.301	2.39	0.010	n.d.	0.5	22.00
27.3	0.41	231.4	367.5	9.2	n.d.	16.4	0.077	0.88	n.d.	0.017	0.58	0.017	n.d.	13.1	3.56
29.3	2.64	260.9	425.2	5.4	n.d.	18.9	0.050	2.13	n.d.	0.024	2.57	0.012	n.d.	—	n.d.
14.7	0.61	142.3	229.0	5.0	n.d.	10.6	0.056	2.27	n.d.	0.017	1.69	0.011	n.d.	17.5	0.74

Common name	Scientific name	Mois-ture (g)	Energy (kcal)	Protein (g)	Fat (g)	Carbo-hydrate (g)	Fiber (g)	Ash (g)	Na (mg)
water snake	*Acrochochordus javanicus*	85.4	67	11.8	1.5	1.6	—	1.0	79.0
sea turtle	*Chelonia mydas*	77.7	98	16.4	2.7	1.9	—	1.3	120.9
tortoise	*Chelodina* sp.	77.2	109	16.3	4.5	0.8	—	1.2	94.8
Insects									
grub (larva)	Family Cerambycidae	55.9	266	20.2	19.6	2.1	—	2.2	7.1
sago grub (larva)	*Rhynchophorus ?schach*	62.9	240	12.1	19.1	4.9	—	1.0	2.8
sago grub (larva)	*Rhynchophorus ?schach*	79.2	126	7.4	9.1	3.5	—	0.9	18.2
tree ant	*Oecophylla* sp.	51.5	122	16.8	4.0	4.8	—	22.9	87.1
tree ant	*Oecophylla virescens*	78.3	111	8.9	5.8	5.8	—	1.3	58.6
Freshwater fish									
jardine's barramundi	*Scleropages leichardti*	77.4	104	16.0	3.6	1.9	—	1.0	54.4
catfish	Family Tachysuridae	80.1	83	16.5	1.6	0.6	—	1.1	57.9
catfish	*Neosilus ?ater*	82.6	74	11.4	2.0	2.5	—	1.5	67.6
gudgeon	*Ophiocra aporos*	78.7	84	17.4	1.0	1.3	—	1.7	45.2
gudgeon	*Eleotris ?fuscus*	81.8	74	14.9	1.2	1.0	—	1.1	33.0
Seawater fish									
barramundi	*Lates calcarifer*	78.0	84	19.4	0.3	1.0	—	1.3	76.3
jardine's barramundi	*Scleropages leichardti*	78.9	87	14.6	1.3	4.1	—	1.1	74.3
shark	?	81.6	82	13.6	1.7	3.1	—	1.5	112.1
shark	?	79.4	79	15.1	0.3	4.0	—	1.2	150.1
catfish	Family Tachysuridae	80.3	81	14.6	1.4	2.4	—	1.3	74.1
threadfin salmon	*Polydactylus plebeius*	77.9	86	16.2	0.5	4.1	—	1.4	71.3
Freshwater shellfish									
mangrove clam	*Gelonia* sp.	87.3	48	6.9	0.7	3.6	—	1.4	101.9
corbicula	Family Corbiculidae	83.1	76	7.5	2.5	5.9	—	0.9	13.7
Seawater shellfish									
clam	Family Veneridae	—	—				—	—	251.8
clam	Family Veneridae	87.4	45	7.1	0.6	2.8	—	2.1	298.9
Freshwater crustaceans									
crayfish	*Cherax communis*	—	—	—	—	—	—	—	97.5
prawn	*Macrobrachium* sp.	82.7	67	13.1	0.5	2.6	—	1.1	63.9
Purchased animal foods									
corned beef	—	58.7	240	21.0	17.3	0.1	—	2.9	1013.1
canned mackerel	—	67.6	176	17.9	11.6	0.1	—	2.8	589.7
canned mackerel	—	67.3	184	18.0	12.4	0.1	—	2.2	304.8
canned cream	—	71.2	115	8.0	1.6	17.2	—	2.0	127.2

— : not measured; n.d.: not detected; the detection limits are

Al: 0.1 mg, V: 20 μg, Cr: 2 μg, Mn: 1 μg, Fe: 5 μg, Ni: 20 μg, Cu: 2 μg, Cd: 5 μg, Hg: 0.05 μg, Pb: 0.5 μg.

a) Detection limit for Cr is not 2 μg but 20 μg, and that for Pb not 0.5 μg but 50 μg.

b) Pumpkin is arbitrarily categorized as a tuber.

c) Abika (or aibika) in Tok Pisin (lingua franca in Papua New Guinea, formerly called Pidgin English).

d) Bedum in Gidra language.

Mg (mg)	Al (mg)	P (mg)	K (mg)	Ca (mg)	V (µg)	Cr (µg)	Mn (mg)	Fe (mg)	Ni (µg)	Cu (mg)	Zn (mg)	Sr (mg)	Cd (µg)	Hg (µg)	Pb (µg)
18.9	0.30	160.0	293.5	9.3	n.d.	6.3	0.057	0.71	n.d.	0.013	1.76	0.015	n.d.	130.6	n.d.
24.1	1.22	156.3	297.1	8.0	n.d.	25.0	0.034	4.89	n.d.	0.073	2.04	0.102	n.d.	0.2	3.28
28.7	2.58	221.0	392.5	12.3	n.d.	49.8	0.038	3.47	n.d.	0.057	1.57	0.037	n.d.	3.8	6.17
21.9	4.63	177.1	242.7	9.2	n.d.	9.9	0.093	5.88	n.d.	0.243	1.59	0.042	n.d.	1.7	n.d.
53.8	1.76	116.6	235.0	17.5	n.d.	14.6	3.891	3.13	n.d.	0.776	1.94	0.083	n.d.	0.6	n.d.
52.8	0.56	101.5	231.2	17.0	n.d.	8.4	0.430	1.97	n.d.	0.637	2.86	0.113	n.d.	—	n.d.
33.9	8.17	250.9	262.5	23.2	n.d.	19.4	4.394	10.56	n.d.	0.424	4.91	0.600	n.d.	0.5	1.80
26.5	3.43	203.1	207.7	17.3	n.d.	4.2	1.367	23.66	n.d.	0.472	3.67	0.115	n.d.	0.6	2.12
29.2	0.53	168.8	356.3	118.3	n.d.	45.0	0.070	0.94	n.d.	0.023	1.53	0.449	n.d.	47.8	4.44
30.2	0.82	203.2	450.0	14.0	n.d.	33.7	0.063	1.26	n.d.	0.016	1.04	0.061	n.d.	24.7	n.d.
24.6	0.96	254.4	319.9	208.9	n.d.	46.0	0.404	1.36	n.d.	0.036	0.84	0.797	n.d.	10.9	n.d.
33.8	0.83	240.0	401.3	71.4	n.d.	15.9	0.073	0.80	n.d.	0.032	1.10	0.217	n.d.	17.1	n.d.
29.6	0.70	177.4	328.4	23.4	n.d.	16.5	0.113	0.65	n.d.	0.014	0.29	0.045	n.d.	22.0	n.d.
35.5	0.64	241.7	445.7	11.9	n.d.	39.1	0.024	2.22	n.d.	0.024	0.50	0.024	n.d.	14.9	n.d.
35.3	0.43	243.6	470.1	8.8	n.d.	26.1	0.017	1.19	n.d.	0.015	0.40	0.034	n.d.	61.7	1.98
33.9	1.14	282.9	417.9	74.4	n.d.	14.6	0.088	3.55	n.d.	0.062	0.46	0.480	n.d.	46.2	2.09
29.0	0.61	226.2	381.8	7.0	n.d.	13.9	0.028	1.48	n.d.	0.007	0.57	0.065	n.d.	186.3	6.93
38.0	0.98	218.3	373.6	24.0	n.d.	37.0	0.054	2.64	n.d.	0.037	0.81	0.137	n.d.	3.8	1.09
34.4	0.14	270.9	486.8	7.4	n.d.	13.8	0.003	0.43	n.d.	0.008	0.40	0.016	n.d.	61.7	1.06
28.4	11.74	159.9	93.2	39.0	40	23.7	3.345	38.44	77	0.279	1.55	0.496	8.8	24.8	4.15
13.4	0.42	194.4	34.0	19.9	n.d.	4.1	0.281	1.36	n.d.	0.097	1.23	0.055	n.d.	2.8	1.15
165.8	20.23	584.9	193.0	533.5	265	274.5	277.151	28.89	152	0.470	1.04	8.352	40.6	0.9	299.40
74.9	111.16	133.0	210.2	47.8	303	112.3	1.549	82.95	88	0.243	1.19	2.083	14.7	1.9	243.30
53.7	0.65	445.5	509.5	24.0	n.d.	5.8	0.160	0.85	n.d.	0.625	2.23	0.080	n.d.	10.2	5.54
28.5	0.14	280.9	311.7	15.8	n.d.	5.5	0.049	0.27	n.d.	0.370	1.11	0.045	n.d.	4.8	n.d.
19.1	n.d.	133.2	175.1	7.7	n.d.	25.3	0.016	5.09	n.d.	0.064	5.49	0.057	n.d.	—	7.70
24.9	n.d.	241.3	241.8	189.0	n.d.	49.1	0.005	2.93	n.d.	0.086	1.41	0.859	n.d.	—	n.d.
31.1	n.d.	188.3	318.8	33.8	n.d.	26.8	0.005	1.18	n.d.	0.070	0.70	0.133	n.d.	—	n.d.
26.3	n.d.	286.3	338.5	345.5	n.d.	66.1	0.022	0.68	n.d.	0.034	1.72	0.255	n.d.	—	7.43

REFERENCES

Akimichi, T. (1984). Children's illness and curing in lowland Papua. *Bull. Natl. Mus. Ethnol.* **9**: 349–382. (in Japanese).

Allaway, W.H. (1986). Soil-plant-animal and human interrelationships in trace element nutrition. In: W. Mertz (ed.), Trace Elements in Human and Animal Nutrition, Vol. 2 (5th ed.). Academic Press, Orlando, pp. 465–488.

Allen, L.H. (1982). Calcium bioavailability and absorption: A review. *Am. J. Clin. Nutr.* **35**: 783–808.

Altshul, A.M. and Grommet, J.K. (1980). Sodium intake and sodium sensitivity. *Nutr. Rev.* **38**: 393–402.

Anderson, R.A. and Kozlovsky, A.D. (1985). Chromium intake, absorption and excretion of subjects consuming self-selected diets. *Am. J. Clin. Nutr.* **41**: 1177–1183.

A.O.A.C. (1970). Official Methods of Analysis of the Association of Official Agricultural Chemists. 11th ed., Association of Official Analytical Chemists, Washington, D.C.

Arroyave, G. and Lee, M. (1966). Variation in urinary excretion of urea and N^1-methylnicotinamide during the day comparison with fasting levels. *Arch. Latinoam. Nutr.* **16**: 125–130.

Ashizawa, K., Kusumoto, A. and Kawabe, T. (1988). Height, weight and chest circumference in a group of Philippine rural children compared with the Japanese standard. *J. Hum. Ergol.* **16**: 43–53.

Bailey, K.V. (1963). Nutritional status of east New Guinean populations. *Trop. Geogr. Med.* **15**: 389–402

Bailey, K.V. (1964a). Growth of Chimbu infants in the New Guinea Highlands. *J. Trop. Pediatr. Afr. Child Health* **10**: 3–16.

Bailey, K.V. (1964b). Dental development in New Guinean infants. *J. Pediatr.* **64**: 97–100.

Bailey, K.V. (1968). Composition of New Guinea highland foods. *Trop. Geogr. Med.* **20**: 141–146.

Barker, D.K. (1965). A study of the eruption times of the deciduous and permanent dentitions of the children of the Territory of Papua and New Guinea. Report submitted to the Department of Public Health, Port Moresby.

Barnes, R. (1963). A comparison of growth curves of infants from two weeks to 20 months in various areas of the Chimbu Subdistrict of the Eastern Highland of New Guinea. *Med. J. Aust.* **2**: 262–266.

Barnstein, F. (1900). Über eine Modifikation des von Ritthausen vorge-schlogenen Verfahrens zur Eiweissbestimmung. *Landw. Vers.-Stat.* **24**: 327–336.

Barrau, J. (1959). The Sago palms and other food plants of marsh dwellers in the Pacific Islands. *Econ. Bot.* **13**: 151–162.

Bayliss-Smith, T.P. (1977). Human ecology and island populations: The problem of change. In: T.B. Bayliss-Smith and R. Feachem (eds.), Subsistence and Survival: Rural Ecology in the Pacific. Academic Press, London, pp. 11–20.

Beaver, W.N. (1920). Unexplored New Guinea. Seely, Service & Co., Edinburgh.

Becroft, C.T. (1967a). Child-rearing practices in the Highlands of New Guinea: A longitudinal study of breast feeding. *Med. J. Aust.* **2**: 598–601.

Becroft, C.T. (1967b). Child-rearing practices in the Highlands of New Guinea: General features. *Med. J. Aust.* **2**: 810–813.

Bentley, G.R. (1985). Hunter-gatherer energetics and fertility: A reassessment of the !Kung San. *Hum. Ecol.* **13**: 79–109.

Berkey, C.S., Read, R.B. and Valadian, I. (1983). Midgrowth spurt in height of Boston children. *Ann. Hum. Biol.* **10**: 25–30.

Bhat, K.R., Arunachalam, J., Yegnasubramanian, S. and Gangadharan, S. (1982). Trace elements in hair and environmental exposure. *Sci. Total Environ.* **22**: 169–178.

Bicchieri, M.G. (ed.) (1972). Hunters and Gatherers Today. Holt, Rinehart and Winston, New York.

Bindra, G.S. and Gibson, R.S. (1986). Iron status of predominantly lacto-ovo vegetarian. East Indian immigrants to Canada: A model approach. *Am. J. Clin. Nutr.* **44**: 643–652.

Birdsell, J.B. (1968). Some predictions for the Pleistocene based on equilibrium systems among recent hunter-gatherers. In: R.B. Lee. and I. DeVore (eds.), Man the Hunter. Aldine, Chicago, pp. 229–240.

Birdsell, J.B. (1973). A basic demographic unit. *Curr. Anthropol.* **14**: 337–356.

Bock, R.D., Wainer, H., Peterson, A., Thissen, D., Murray, J. and Roche, A.F. (1973). A parameterization for individual human growth curve. *Hum. Biol.* **45**: 63–80.

Bodmer, W.F. and Cavalli-Sforza, L.L. (1968). A migration matrix model for the study of random genetic drift. *Genetics* **59**: 565–592.

Bodmer, W.F. and Cavalli-Sforza, L.L. (1974). The analysis of genetic variation using migration matrics. In: J.F. Crow and C. Denniston (eds.), Genetic Distance. Plenum, New York, pp. 45–61.

Bohdal, M. and Simmons, W.K. (1968). A comparison of the nutritional indices in healthy African, Asian, and European children. *Bull. WHO* **40**: 166–170.

Bourke, R.M. (1982). Root crops in Papua New Guinea. In: R.M. Bourke and V. Kasavan (eds.), Proceedings of Second Papua New Guinea Food Crops Conference. Department of Primary Industry, Port Moresby.

Boyce, A.J., Attenborough, R.O. and Harrison, G.A. (1978). Variation in blood pressure in a New Guinea population. *Ann. Hum. Biol.* **5**: 313–319.

Bradfield, R.B., Huntzicker, P.B. and Fruehan, G.J. (1969). Simultaneous comparison of respirometer and heart-rate telemetry technique as measures of human energy expenditure. *Am. J. Clin. Nutr.* **22**: 696–700.

Brainard, J.M. (1982). Mating structure of a nomadic pastoral population. *Hum. Biol.* **54**: 469–475.

Bray, R.S. (1979). Insect-borne diseases: Malaria. In: R. Chambers, R. Longhurst and A. Pagey (eds.), Seasonal Dimensions to Rural Poverty. Frances Pinter, Allanheld, pp. 116–120.

Brennan, E.R. (1981). Kinship, demographic, social, and geographic characteristics of mate choice in Sanday, Orkney Islands, Scotland. *Am. J. Phys. Anthropol.* **55**: 129–138.

Brennan, E.R. and Boyce, A.J. (1980). Mate choice and marriage on Sanday,

Orkney Islands. In: B. Dyke and W.T. Morrill (eds.), Genealogical Demography. Academic Press, New York, pp. 197–207.

Brožek, J., Grande, F., Anderson, J.T. and Keys, A. (1963). Densitometric analysis of body composition: Revision of quantitative assumptions. *Ann. N.Y. Acad. Sci.* **110**: 113–140.

Buchet, J.P., Lauwery, R., Vandevoorde, A. and Pycke, J.M. (1983). Oral daily intake of cadmium, lead, manganese, copper, chromium, mercury, calcium, zinc and arsenic in Belgium: A duplicate meal study. *Food Chem. Toxic.* **21**: 19–24.

Burman, D. (1982). Nutrition in early childhood. In: D.S. McLaren and D. Burman (eds.), Textbook of Paediatric Nutrition. Churchill Livingstone, Edinburgh, pp. 39–73.

Burns-Cox, C.J. (1970). Splenomegaly and blood pressure in an Orang-Asli community in West Malaysia. *Am. Heart J.* **80**: 718–719.

Buzina, R., Jusic, M., Sapunar, J. and Milanovic, N. (1980). Zinc nutrition and taste acuity in school children with impaired growth. *Am. J. Clin. Nutr.* **33**: 2262–2267.

Byrne, A.R. and Kosta, L. (1979). On the vanadium and tin contents of diet and human blood. *Sci. Total Environ.* **13**: 87–90.

Calabrese, E.J., Moore, G.S., Tuthill, R.W. and Sieger, T.L. (eds.) (1980). Drinking Water and Cardiovascular Diseases. Pathotox, Illinois.

Calkins, B.M., Whittaker, D.J., Nair, P.P., Rider, A.A. and Turjman, N. (1984). Diet, nutrition intake, and metabolism in populations at high and low risk for colon cancer. *Am. J. Clin. Nutr.* **40**: 896–905.

Carneiro, C.R. (1957). Subsistence and Social Structure. Ph.D. Dissertation, University of Michigan.

Carter, C.O. and Marshall, W.A. (1978). The genetics of adult stature. In: F. Falkner and J.M. Tanner (eds.), Human Growth 1: Principles and Prenatal Growth. Plenum, New York, pp. 299–305.

Cassens, R.G., Briskey, E.J. and Hoekstra, W.G. (1963). Variation in zinc content and other properties of various porcine muscles. *J. Sci. Food Agr.* **14**: 427–432.

Cataldo, D.A., Garlando, T.R., Wildung, R.E. and Drucker, H. (1978). Nickel in plants. II. Distribution and chemical form in soybean plants. *Plant Phys.* **62**: 566–570.

Cawte, J. (1984). Emic accounts of a mystery illness: The Groote Eylandt syndrome. *Aust. New Zealand J. Psychiat.* **18**: 179–187.

Champness, L.T., Bradley, M.A. and Walsh, R.J. (1963). A study of the Tolai in New Britain. *Oceania* **34**: 66–75.

Chase, H.P., Hambidge, K.M., Barnett, S.E., Houts-Jakobs, M.J., Lenz, K. and Gillespie, J. (1980). Low vitamin A and zinc concentrations in Mexican-American migrant children with growth retardation. *Am. J. Clin. Nutr.* **33**: 2346–2349.

Chittleborough, G. (1980). A chemist's view of the analysis of human hair for trace elements. *Sci. Total Environ.* **14**: 53–75.

Clarke, D.J. (1978). A Handbook for Field Workers: Food Habits of the People of Papua New Guinea. Mt. Hagen. (Mimeograph).

Clarke, H.H. (1966). Muscular Strength and Endurance in Man (5th ed.). Prentice-Hall, Englewood Cliffs.

Clement, F.J. (1974). Longitudinal and cross-sectional assessments of age changes in physical strength as related to sex, social class, and mental ability. *J. Gerontol.* **29**: 423–429.

Clemente, G.F., Rossi, L.C. and Santaroni, G.P. (1977). Trace element intake and excretion in the Italian population. *J. Radioanal. Chem.* **37**: 549–558.

Cohen, B.L. (1985). Bioaccumulation factors in marine organisms. *Health Phys.* **49**: 1290–1294.

Conklin, H.C. (1957). Hanunóo Agriculture. FAO, Rome.

Cook, J., Altman, D.G., Moore, D.M.C., Topp, S.G., Holland, W.W. and Elliot, A. (1973). A Survey of the nutritional status of school children: Relation between nutrient intake and socio-economic factors. *Br. J. Prevent. Soc. Med.* **27**: 91–99.

Coon, C. (1954). Climate and Race. Smithsonian Report for 1953. Smithsonian Institution, Washington, D.C., pp. 277–298.

Corden, M.W. (1970). Some observation on village life in New Guinea. *Aust. Inst. Anat. Food Nutr. Notes and Reviews* **27**: 77–82.

Count, E.W. (1942). A quantitative analysis of growth in certain human skull dimensions. *Hum. Biol.* **14**: 143–165.

Count, E.W. (1943). Growth patterns of the human physique: An approach to kinetic anthropometry. Part I. *Hum. Biol.* **15**: 1–32.

Covee, L.M.J., Nugteren, D.H. and Luyken, R. (1962). The nutritional condition of the Kapaukus in the Central Highlands of Netherlands New Guinea. I. Biochemical examinations. *Trop. Geogr. Med.* **14**: 27–35.

Cowgill, U.M. (1981). The chemical composition of bananas: Market basket values, 1968-1980. *Biol. Trace Element Res.* **3**: 33–54.

Cresta, M., Allegrini, M., Casadei, E., Gallorini, M., Lanzola, E. and Panatta, G.B. (1976). Benin: nutritional considerations on trace elements in diet. *Food Nutr.* **2**: 8–18.

Crittenden, R. and Baines, J. (1986). The seasonal factors influencing child malnutrition on the Nembi Plateau, Papua New Guinea. *Hum. Ecol.* **14**: 191–223.

Crounse, R.G., Pories, W.J., Bray, J.T. and Mauger, R.L. (1983a). Geochemistry and man: Health and disease. 1. Essential elements. In: I. Thornton (ed.), Applied Environmental Geochemistry. Academic Press, London, pp. 267–308.

Crounse, R.G., Pories, W.J., Bray, J.T. and Mauger, R.L. (1983b). Geochemistry and man: health and disease. 2. Elements possibly essential, those toxic and others. In: I. Thornton (ed.). Applied Environmental Geochemistry. Academic Press, London, pp. 309–333.

Dahl, L.K. (1972). Salt and hypertension. *Am. J. Clin. Nutr.* **25**: 231–244.

Damon, A. (1968). Secular trend in height and weight within old American families at Harvard, 1870-1965. *Am. J. Phys. Anthropol.* **29**: 45–50.

Daru General Hospital (1981). Epidemiological Report for 1980-1981. Division of Health, Malaria Service of the Daru General Hospital, Daru. (Mimeograph).

Deming, J. (1957). Application of the Gompertz curve to the observed pattern of growth in length of 48 individual boys and girls during the adolescent cycle of growth. *Hum. Biol.* **29**: 83–122.

Deming, J. and Washburn, A.H. (1963). Application of the Jenss curve to the observed pattern of growth during the first eight years of life in forty boys and forty girls. *Hum. Biol.* **35**: 484–506.

Dennet, G. and Connell, J. (1988). Acculturation and health in the Highlands of Papua New Guinea: dissent on diversity, diets and development. *Curr. Anthropol.* **29**: 273–299.

Di Giulio, R.T. and Scanlon, P.F. (1985). Heavy metals in aquatic plants, clams, and sediments from the Chesapesake Bay, U.S.A.: Implications for waterfowl. *Sci. Total Environ.* **41**: 259–274.

Dixon, W.J. and Brown, M.B. (1979). BMDP Biomedical Computer Programs, P-Series. University of California Press, Berkeley.

Dobzhansky, T. (1968). Adaptedness and fitness. In: R.C. Lewontin (ed.), Population Biology and Evolution. Syracuse University Press, New York, pp. 109–121.

Doi, R., Raghupathy, L., Ohno, H., Naganuma, A., Imura, N. and Harada, M. (1988). A study of the sources of external metal contamination of hair. *Sci. Total Environ.* **77**: 153–161.

Downs, M.C. (1972). Deer. In: P. Ryan (ed.), Encyclopaedia of Papua and New Guinea. Melbourne University Press, Melbourne, pp. 240–241.

Drews. L.M., Kies, C. and Fox, H.M. (1979). Effect of dietary fiber on copper, zinc, and magnesium utilization by adoelscent boys. *Am. J. Clin. Nutr.* **32**: 1893–1897.

Duncan, D.B. (1975). t tests and intervals for comparisons suggested by the data. *Biometrics* **31**: 339–359.

Durnin, J.V.G.A. and Womersley, J. (1974). Body fat assessed from total body density and its estimation from skinfold thickness: Measurements on 481 men and women aged from 16 to 72 years. *Br. J. Nutr.* **32**: 77–97.

Dwyer, P. (1983). Etolo hunting performance and energetics. *Hum. Ecol.* **11**: 145–174.

Dyke, B. (1971). Potential mates in a small human population. *Soc. Biol.* **18**: 28–39.

Dyke, B. (1984). Migration and the structure of small populations. In: A. J. Boyce (ed.), Migration and Mobility: Biosocial Aspects of Human Movement. Talylor & Francis, London, pp. 69–81.

Early, J.D. (1985). Low forager fertility: Demographic characterisitc or methodological artifact? *Hum. Biol.* **57**: 387–99.

Eisenberg, M. and Topping, J.J. (1984). Trace metal residues in shellfish from Maryland waters, 1976-1980. *J. Environ. Sci. Health* B**19**: 649–671.

El Lozy, M. (1978). A critical analysis of the double and triple logistic growth curves. *Ann. Hum. Biol.* **5**: 389–394.

Ellen, R. (1982). Environment, Subsistence and System: The Ecology of Small-Scale Formations. Cambridge University Press, Cambridge.

Eminians, J., Reinhold, J.G., Kfoury, G.A., Amirhakimi, G.H., Sharif, H. and Ziai, M. (1967). Zinc nutrition of children in Fars Province of Iran. *Am. J. Clin. Nutr.* **20**: 734–742.

Erasmus, C.J. (1955). Work pattern in a Mayo village. *Am. Anthropol.* **57**: 322–333.

Evans, W.E., Read, J.I. and Caughlin, D. (1985). Quantification of results for estimating elemental dietary intakes of lithium, rubidium, strontium, molybdenum, vanadium and silver. *Analyst* **110**: 873–877.

Eveleth, P.B. and Tanner, J.M. (1976). Worldwide Variation in Human Growth. Cambridge University Press, Cambridge.

Ewers, W.H. and Jeffrey, W.T. (1971). Parasites of Man in Niugini. The

Jacaranda Press, Milton, pp. 137–194.

FAO (1979). The Economic Value of Breast-Feeding. (FAO Food and Nutrition Paper 11). Food and Agriculture Organization of the United Nations, Rome.

FAO/USDHEW (1968). Food Composition Table for Use in Africa. Food and Agriculture Organization of the United Nations, Rome.

FAO/USDHEW (1972). Food Composition Table for Use in East Asia. Food and Agriculture Organization of the United Nations, Rome.

FAO/WHO (1973). Energy and Protein Requirements. Food and Agriculture Organization of the United Nations, Rome.

Farnworth, E.R. (1974). Sodium, potassium, calcium and magnesium in foods collected at Koki market: Part 1. *Sci. New Guinea* **2**: 182–186.

Farre, R. and Lagarda, M.J. (1986). Chromium contents of foods and diets in a Spanish population. *J. Micronutr. Anal.* **2**: 297–304.

Favretto, L. and Favretto, L.G. (1984a). Multivariate data analysis of some xenobiotic trace metals in mussels from the Gulf of Trieste. *Z. Lebensm. Unters. Forsch.* **179**: 201–204.

Favretto, L. and Favretto, L.G. (1984b). Principal component analysis as a tool for studying interdependences among trace metals in edible mussels from the Gulf of Trieste. *Z. Lebensm. Unters. Forsch.* **179**: 377–380.

Fergusson, J.E., Holzbecher, J. and Ryan, D.E. (1983). The sorption of copper (II), manganese (II), zinc (II) and arsenic (III) onto human hair, and their desorption. *Sci. Total Environ.* **26**: 121–135.

Ferro-Luzzi, A., Norgan, N.G. and Durnin, J.V.G.A. (1975). Food intake, its relationship to body weight and age, and its apparent nutritional adequency in New Guinea children. *Am. J. Clin. Nutr.* **28**: 1443-1453.

Ferro-Luzzi, A., Norgan, N.G. and Durnin, J.V.G.A. (1978). The nutritional status of some New Guinea children as assessed by anthropometric, biochemical and other indices. *Ecol. Food Nutr.* **7**: 115–128.

Fisher, R.A. (1958). The Genetical Theory of Natural Selection. Dover, New York.

Fleischmann, L. and Turpeinen, S. (1976). A dialect survey of Eastern Trans-Fly Languages. *Pacific Linguistics* A**45**: 39–76.

Flyvholm, M-A., Nielsen, G.D. and Andersen, A. (1984). Nickel content of food and estimation of dietary intake. *Z. Lebensm. Unters. Forsch.* **179**, 427–431.

Forbes, G.B. and Bruining, G.J. (1976). Urinary creatine excretion and lean body mass. *Am. J. Clin. Nutr.* **29**: 1359–1366.

Fountain, O.C. (1966). Wulkum: Land, Livelihood and Change in a New Guinea Village. M.A. Thesis, Victoria University of Wellington, New Zealand.

Fransson, G-B., Gebre-Medhin, M. and Hambraeus, L. (1984). The human milk contents of iron, copper, zinc, calcium and magnesium in a population with a habitually high intake of iron. *Acta Pediatr. Scand.* **73**: 471–476.

Freedman, L. and Macintosh, N.W.G. (1965). Stature variation in Western Highland males of east New Guinea. *Oceania* **35**: 286–304.

Freeland-Graves, J.H., Bodzy, P.W. and Eppright, M.A. (1980). Zinc status of vegetarians. *J. Am. Diet. Assoc.* **77**: 655–661.

Freis, E. (1976). Salt, volume and the prevention of hypertension. *Circulation* **53**: 589–595.

Friedlaender, J.S. (1975). Patterns of Human Variation: The Demography, Genetics and Phenetics of Bougainville Islanders. Harvard University Press, Cambridge.

Frisancho, A.R. (1981). New norms of upper limb fat muscle areas for assessment of nutritional status. Am. J. Clin. Nutr. 34: 2540–2545.

Frisancho, A.R., Cole, P.E. and Klayman, J.E. (1977). Greater contribution to secular trend among offspring of short parents. Hum. Biol. 49: 51–60.

Fujita, Y., Okuda, T., Rikimaru, T., Ichikawa, M., Miyatani, S., Kajiwara, N.M., Date, C., Minamida, T., Koishi, H., Alpers, M.P. and Heywood, P.F. (1986). Studies of nitrogen balance in male Highlanders in Papua New Guinea. J. Nutr. 116: 536–544.

Fukushima, I., Imahori, A., Shiobara, S., Ebato, K. and Hyoi, N. (1982). A method to monitor trace elements in the community environment through the chemical analysis of hair. Jpn. J. Hyg. 37: 768–778.

Gasser, T., Müller, H., Köhler, W., Prader, A., Largo, R. and Molinari, L. (1985). An analysis of the mid-growth and adolescent spurts of height based on accelaration. Ann. Hum. Biol. 12: 129–148.

Gebre-Medhin, M. and Birgegard, G. (1981). Serum ferritin in Ethiopian mothers and their newborn infants: Relation to iron intake and socio-economic conditions. Scand. J. Haematol. 27: 247–252.

Gebre-Medhin, M., Killander, A., Vahlqvist, B. and Wuhib, E. (1976). Rarity of anaemia of pregnancy in Ethiopia. Scand. J. Haematol. 16: 168–175.

Genest, P.E. and Hatch, W.I. (1981). Heavy metals in Mercenaria mercenaria and sediments from the New Bedford Harbor region of Buzzard's Bay, Massachusetts. Bull. Environ. Contam. Toxicol. 26: 124–130.

Gentile, P.S., Trentalange, M.J. and Coleman, M. (1981). The relationship of hair zinc concentrations to height, weight, age, and sex in the normal populations. Pediatr. Res. 15: 123–127.

Gibson, R.S., Anderson, B.M. and Sabry, J.H. (1983). The trace metal status of a group of post-menopausal vegetarians. J. Am. Diet. Assoc. 82: 246–250.

Golson, J. (1977). No room at the top: agricultural intensification in the New Guinea Highlands. In: J. Allen, J. Golson and R. Jones (eds.), Sunda and Sahul: Prehistoric Studies in Southeast Asia, Melanesia and Australia. Academic Press, London, pp. 601–638.

Gormican, A. (1970). Inorganic elements in foods used in hospital menus. J. Am. Diet. Assoc. 56: 397–403.

Goto, S., Sawamura, T. and Seki, H. (1972). Mineral contents in Japanese daily diet, Ca, P, Mg, K and Na. J. Jpn. Soc. Food Nutr. 25: 359–361. (in Japanese with English abstract).

Gould, S. J. (1966). Allometry and size in ontogeny and phylogeny. Biol. Rev. 41: 587–640.

Greger, J. L. (1985). Aluminum content of the American diet. Food Tech. 39: 73–80.

Greulich, W.W. (1957). A comparison of the physical growth and development of American-born and native Japanese. Am. J. Phys. Anthropol. 15: 489–515.

Greulich, W.W. (1976). Some secular changes in the growth of American-born and native Japanese children. Am. J. Phys. Anthropol. 45: 553–568.

Gross, D.R. (1984). Time allocation: A tool for the study of cultural behavior.

Ann. Rev. Anthropol. **13**: 519–558.

Hallberg, L. (1981). Bioavailability of dietary iron in man. *Ann. Rev. Nutr.* **1**: 123–147.

Hallberg, L., Brune, M. and Rossander, L. (1989). Iron absorption in man: Ascorbic acid and dose-dependent inhibition by phytate. *Am. J. Clin. Nutr.* **49**: 140–144.

Hambidge, K.M. (1973). Increase in hair copper concentration with increasing distance from the scalp. *Am. J. Clin. Nutr.* **26**: 1212–1215.

Hambidge, K.M., Casey, C.E. and Krebs, N.F. (1986). Zinc. In: W. Mertz (ed.), Trace Elements in Human and Animal Nutrition, Vol. 2 (5th ed.). Academic Press, Orlando, pp. 1–137.

Hambidge, K.M., Hambidge, C., Jakobs, M. and Baum, J.D. (1972). Low levels of zinc in hair, anorexia, poor growth, and hypogeusia in children. *Pediatr. Res.* **6**: 868–874.

Hambidge, K.M., Walravens, P.A., Brown, R.M., Webster, J., White, S., Anthony, M. and Roth, M.L. (1976). Zinc nutrition of preschool children in the Denver Head Start program. *Am. J. Clin. Nutr.* **29**: 734–738.

Hames, R.B. (1979). A comparison of the efficiencies of the shotgun and the bow in neotropical forest hunting. *Hum. Ecol.* **7**: 219–252.

Hames, R. B. and Vickers, W. T. (eds.) (1983). Adaptive Responses of Native Amazonians. Academic Press, New York.

Hamil, P.V.V., Drizd, T.A., Johnson, C.L., Reed, R.B., Roche, A.F. and Moore, W.M. (1979). Physical growth: National Center for Health Statistics percentiles. *Am. J. Clin. Nutr.* **32**: 607–629.

Hamilton, E. I. and Minski, M.J. (1972/73). Abundance of the chemical elements in man's diet and possible relations with environmental factors. *Sci. Total Environ.* **1**: 375–394.

Hamilton, M., Pickering, G.W., Roberts, J.A.F. and Sowry, G.S.C. (1954). The aetiology of essential hypertension: The arterial pressure in the general population. *Clin. Sci.* **13**: 11–35.

Harpending, H. and Jenkins, T. (1974). !Kung population structure. In: J.F. Crow and C. Denniston (eds.), Genetic Distance. Plenum, New York, pp. 137–165.

Harris, D.R. (1977). Subsistence strategies across Torres Strait. In: J. Allen, J. Golson and R. Jones (eds.), Sunda and Sahul: Prehistoric Studies in Southeast Asia, Melanesia and Australia. Academic Press, London, pp. 421–462.

Harris, J.E., Fabris, G.J., Statham, P.J. and Tawfik, F. (1979). Biogeochemistry of selected heavy metals in Western Port, Victoria, and use of invertebrates as indicators with emphasis on *Mytilus edulis planulatus*. *Aust. J. Mar. Freshwater Res.* **30**: 159–178.

Harrison, G.A. (1977). Introduction: Structure and function in the biology of human populations. In: G.A. Harrison (ed.), Population Structure and Human Variation. Cambridge University Press, Cambridge.

Harvey, P.W. and Heywood, P.F. (1983). Twenty-five years of dietary change in Simbu Province, Papua New Guinea. *Ecol. Food Nutr.* **13**: 27–35.

Harvey, R.G. (1974). An anthropometric survey of growth and physique of the populations of Karkar Island and Lufa Subdistrict, New Guinea. *Phil. Trans. Royal Soc. London* B. **268**: 279–292.

Hashimoto, S. (1983). On Daru, the provincial capital. In: G. Ohshima (ed.),

Peoples of the Torres Strait: Geographical and Ethnological Studies. Kokin-shoin, Tokyo, pp. 231–248. (in Japanese).

Hassan, F.A. (1973). On mechanism of population growth during the Neolithic. *Curr. Anthropol.* **14**: 535–42.

Hassan, F.A. (1981). Demographic Archeology. Academic Press, New York.

Hauspie, R.C., Wachholder, A., Baron, G., Cantraine, F., Susanne, C. and Graffer, M. (1980). A comparative study of the fit of four different functions to longitudinal data of growth in height of Belgian girls. *Ann. Hum. Biol.* **7**: 347–358.

Hazell, T. (1985). Minerals in foods: Dietary sources, chemical forms, interactions, bioavailability. *Wld. Rev. Nutr. Diet.* **46**: 1–123.

Heath, B.H. and Carter, J.E.L. (1971). Growth and somatotype patterns of Manus children, Territory of Papua and New Guinea: Application of a modified somatotype method to the study of growth patterns. *Am. J. Phys. Anthropol.* **35**: 49–68.

Heymsfield, S.B., Arteaga, C., McManus, C., Smith, J. and Moffitt, S. (1983). Measurement of muscle mass in humans: Validity of the 24-hours urinary method. *Am. J. Clin. Nutr.* **37**: 478–494.

Heywood, P.F. (1983). Growth and nutrition in Papua New Guinea. *J. Hum. Evol.* **12**: 133–143.

Hiernaux, J. (1963). Heredity and environment: Their influence on human morphology: A comparison of two independent lines of study. *Am. J. Phys. Anthropol.* **21**: 575–590.

Hiernaux, J. (1964). Weight/height relationship during growth in Africans and Europeans. *Hum. Biol.* **36**: 273–293.

Hiernaux, J. (1968). Bodily shape differentiation of ethnic groups and of the sexes through growth. *Hum. Biol.* **40**: 44–62.

Hipsley, E.H. and Clements, F.W. (eds.) (1950). Report of the New Guinea Nutrition Survey Expedition, 1947. Department of External Territories of Australia, Canberra.

Hipsley, E.H. and Kirk, N.E. (1965). Studies of dietary intake and the expenditure of energy by New Guineans. Technical Paper No. 147, South Pacific Comission, Noumea.

Hodges, K., Fysh, C.F. and Rienits, K.G. (1950). New Guinea and Papuan food composition tables. In: E.H. Hipsley and F.W. Clements (eds.), Report of the New Guinea Nutrition Survey Expedition, 1947. Department of External Territories of Australia, Canberra, pp. 273–279.

Hofvander, Y. (1968). Hematological investigations in Ethiopia, with special reference to a high iron intake. *Acta Med. Scand.* Suppl. **494**: 1-74.

Holden, J.M., Wolf, W.R. and Mertz, W. (1979). Zinc and copper in self-selected diets. *J. Am. Diet. Assoc.* **75**: 23–28.

Hongo, T., Suzuki, T., Ishida, H. and Suzuki, H. (1988). Variation of zinc concentration with the length of scalp hair in young adult women. *J. Jpn. Soc. Nutr. Food Sci.* **41**: 17–22. (in Japanese with English abstract).

Hope, G.S., Golson, J. and Allen, J. (1983). Palaeoecology and prehistory in New Guinea. *J. Hum. Evol.* **12**: 39–60.

Horiguchi, S., Teramoto, K., Kurono, T. and Ninomiya, K. (1978). The arsenic, copper, lead, manganese and zinc contents of daily foods and beverages in Japan and estimate of their daily intake. *Osaka City Med. J.* **24**: 131–141.

Hoshi, H. (1978). Allometric analysis on growth of Japanese-American hybrids with special regard to somatotypic relation between hybrids and their parental populations. *Acta Anat. Nippon,* **53**: 283–296. (in Japanese with English summary).

Howell, N. (1979). Demography of Dobe !Kung. Academic Press, New York.

Hudson, G.J., John, P.M.V. and Paul, A.A. (1980). Variation in the composition of Gambian foods: The importance of water in relation to energy and protein content. *Ecol. Food Nutr.* **10**: 9–17.

Huxley, J.S. (1932). Problems of Relative Growth. Mathuen, London.

Huxley, J.S. and Teissier, G. (1936). Terminology of relative growth. *Nature* **137**: 780–781.

Hyndman, D.C., Ulijaszek, S.J. and Lourie, J.A. (1989). Variability in body physique, ecology and subsistence in the Fly River region of Papua New Guinea. *Am. J. Phys. Anthropol.* **79**: 81–88.

Imahori, A., Fukushima, I., Shiobara, S., Yanagida, Y. and Tomura, K. (1979). Multielement neutron activation analysis of human scalp hair: A local population survey in the Tokyo Metropolitan area. *J. Radioanal. Chem.* **52**: 167–180.

Inaoka, T. (in press). Energy expenditure of the Gidra in lowland Papua: the validity of heart rate method in the field. *Man Cul. Oceania* **6**.

Inoue, Y., Morino, M., Kawada, J., Fukushima, M., Saito, Y., Kagawa, Y., Ishii, K. and Shibata, H. (1985). Correlation of Na, K and Na/K ratio estimated from dietary records with 24 hours urine samples. *J. Jpn. Soc. Nutr. Food Sci.* **38**: 259–264. (in Japanese with English abstract).

Ishida, H., Hongo, T., Ohba, T., Suzuki, H. and Suzuki, T. (1988). Dietary zinc intake and plasma and urinary zinc levels in a group of young adult women. *J. Jpn. Soc. Nutr. Food Sci.* **41**: 373–380. (in Japanese with English abstract).

Ishigure, K., Ikeda, J. and Nagata, H. (1985). Daily variations of energy and nutrient intakes during fifteen months. *J. Jpn. Soc. Nutr. Food Sci.* **38**: 459–464. (in Japanese with English abstract).

Ishimatsu, S. (1988). Nickel contents in various Japanese foodstuffs and estimation of daily intake. *J. Jpn. Soc. Nutr. Food Sci.* **41**: 227-233. (in Japanese with English abstract).

Israelsohn, W.J. (1960). Description and modes of analysis of human growth. In: J.M. Tanner (ed.), Human Growth. Pergamon, New York, pp. 21-42.

Iwao, S. (1977). Cadmium, lead, copper and zinc in food, faces and organs of humans: Interrelationships in food and feces and interactions in the liver and the renal cortex. *Keio J. Med.* **26**:63–78.

Iwao, S., Sugita, M. and Tsuchiya, K. (1981). Some metabolic interrelationships among cadmium, lead, copper and zinc: Results from a field survey in Cd-polluted area in Japan. Part 1: Dietary intake of the heavy metals. *Keio J. Med.* **30**: 17–36.

Jamison, P.L. and Zegura, S.L. (1974). A univariate and multivariate examination of measurement error in anthropometry. *Am. J. Phys. Anthropol.* **40**: 197–204.

Japanese Ministry of Education, Science and Culture (1985). Reports of Physical and Motor Capability Survey in 1984. (Mimeograph) (in Japanese).

Japanese Ministry of Health and Welfare (1984). The Japanese Recommended

Dietary Allowances. Dai-ichi Shuppan, Tokyo. (in Japanese).

Jelliffe, D.B. (1955). Infant nutrition in the subtropics and tropics. WHO Monograph Series 29, Geneva.

Jelliffe, D.B. (1966). The Assessment of the Nutritional Status of the Community. WHO Monograph Series 53, Geneva.

Jelliffe, D.B. and Maddocks, I. (1964). Notes on ecologic malnutrition in the New Guinea Highlands. *Clin. Pediatr.* **3**: 432–438.

Jenkins, C.L., Orr-Ewing, A.K. and Heywood, P.F. (1984). Cultural aspects of early childhood growth and nutrition among the Amele of lowland Papua New Guinea. *Ecol. Food Nutr.* **14**: 261–275.

Jenss, R. M. and Bayley, N. (1937). A mathematical method for studying the growth of a child. *Hum. Biol.* **9**: 556–563.

Johnson, A. (1975). Time allocation in a Machiguenga community. *Ethnology* **14**: 301–310.

Johnston, F.E. (1982). Physical Anthropology. Wm. C. Brown, Dubuque, Iowa.

Kalkwarf, H.J., Haas, J.D., Belko, A.Z., Roach, R.C. and Roe, D.A. (1989). Accuracy of heart-rate monitoring and activity diaries for estimating energy expenditure. *Am. J. Clin. Nutr.* **49**: 37–43.

Kalmus, H. (1969). Ethnic differences in sensory perception. *J. Biosoc. Sci.* Suppl. **1**: 81-90.

Kamakura, M. (1983). A study of the characteristics of trace elements in the hair of Japanese: Reference values and the trace elements patterns for determining normal levels. *Jpn. J. Hyg.* **38**: 823–838.

Kanazawa, H. and Muto, S. (1984). Sodium and potassium intakes through habitual diet and its urinary excretion. *J. Jpn. Soc. Nutr. Food Sci.* **37**: 165–170. (in Japanese with English abstract).

Kannel, W.B., Brand, N., Skinner, J.J. Jr., Dawber, T.R. and McNamara, P.M. (1967). The relation of adiposity to blood pressure and development of hypertension. *Ann. Int. Med.* **67**: 48–59.

Kawabe, T. (1983). Development of hunting and fishing skill among boys of the Gidra in lowland Papua New Guinea. *J. Hum. Ergol.* **12**: 65–74.

Kawabe, T. (1986a). Intrapopulation variation of body physique in the Gidra, Papua New Guinea. *Man Cul. Oceania* **2**: 27–55.

Kawabe, T. (1986b). Ecological Diversity of Growth and Physique of the Gidra in Lowland Papua by Applying a New Method for Growth Curve Estimation. Ph.D. Dissertation, University of Tokyo.

Kawabe, T. (1988). A new application of the two logistic growth curves in stature. *J. Anthropol. Soc. Nippon* **96**: 241.

Kelsay, J.L., Frazier, C.W., Prather, E.S., Canary, J.J., Clark, W.M. and Powell, A.S. (1988). Impact of variation in carbohydrate intake on mineral utilization by vegetarians. *Am. J. Clin. Nutr.* **48**: 875–879.

Keys, A., Fidanza, F., Karvonen, M.J., Kimura, N. and Taylor, H.L. (1972). Indices of relative weight and obesity. *J. Chron. Dis.* **25**: 329–343.

Khaw, K.T. and Barrett, C.F. (1984). Dietary potassium and blood pressure in a population. *Am. J. Clin. Nutr.* **39**: 963–968.

Kiloh, L.G., Lethlean, A.K., Morgan, G., Cawte, J.E. and Harris, M. (1980). An endemic neurological disorder in tribal Australian Aborigines. *J. Neurol. Neurosurg. Psychiatry* **43**: 661–668.

Kimura, K. (1967). A consideration of the secular trend in Japanese for height and weight by a graphic method. *Am. J. Phys. Anthropol.* **27**: 89–94.

Kimura, K. (1970). A study on the variation of individual growth by the relative growth of body weight to height. *Bull. Fac. Phys. Educ. Tokyo Univ. Educ.* **9**: 77–88. (in Japanese with English summary).

Kimura, K. (1977). Has the secular trend for greater height ceased in Japanese? *J. Natl. Def. Med. Coll.* **2**: 72–76. (in Japanese with English summary).

Klaus, M., Dowling, S. and Kennell, J. (1981). Feeding and behavior: Three recent observations. In: J.T. Bond *et al.* (eds.), Infant and Child Feeding. Academic Press, New York, pp. 427–435.

Kobayashi, N. and Brazelton, T.B. (eds.) (1984). The Growing Child in Family and Society: An Interdisciplinary Study in Parent-Infant Bonding. University of Tokyo Press, Tokyo.

Konner, M. and Worthman, C. (1980). Nursing frequency, gonadal function and birth-spacing among !Kung hunter-gatherers. *Science* **207**: 788–791.

Kotake, Y., Ikeda, S. and Shibata, M. (1981). Studies on ingested zinc quantity of Japanese. *J. Jpn. Soc. Nutr. Food Sci.* **34**: 355–365. (in Japanese with English abstract).

Kouchi, M. and Hanihara, K. (1981). An analysis of errors in somatometric research. *J. Anthropol. Soc. Nippon* **89**: 493–504.

Kuchikura, Y. (1988). Food use and nutrition in a hunting and gathering community in transition, Peninsular Malaysia. *Man Cul. Oceania* **4**: 1–30.

Kunisaki, N., Takada, K., Shirano, Y., Asakusa, S., Hanaoka, H. and Matsuura, H. (1984). An investigation into the actual measurement of energy intake, mineral intake and fatty acid composition in hospital diet. *J. Jpn. Soc. Nutr. Food Sci.* **37**: 85–97. (in Japanese with English abstract).

Kyle, J.H. (1981). Mercury in the People and the Fish of the Fly and Strickland River Catchments. Office of Environment and Conservation, Waigani (Port Moresby).

Kyle, J.H. and Ghani, N. (1982). Methylmercury in human hair: A study of a Papua New Guinean population exposed to methylmercury through fish consumption. *Arch. Environ. Health* **37**: 266–70.

Kyle, J.H. and Ghani, N. (1983). Mercury concentrations in canned and fresh fish and its accumulation in a population of Port Moresby residents. *Sci. Total Environ.* **26**: 157–162.

Laird, A.K. (1967). Evolution of the human growth curve. *Growth* **31**: 345–355.

Lambert, J.N. (1980). Bottle-feeding legislation in Papua New Guinea. *J. Hum. Nutr.* **34**: 23–25.

Lambert, J.N. and Basford, J. (1977). Port Moresby infant feeding survey. *Papua New Guinea Med. J.* **20**: 175–181.

Landtman, G. (1927). The Kiwai Papuans of British New Guinea. Methuen, London.

Langley, D. (1950). Food comsumption and dietary levels. In: E.H. Hipsley and F.W. Clements (eds.), Report of the New Guinea Nutrition Survey Expedition, 1947. Department of External Territories of Australia, Canberra, pp. 93–142.

Laughlin, H. (1968). Hunting; An integrating biobehavior system and its evolutionary importance. In: R.B. Lee and I. DeVore (eds.), Man the Hunter. Aldine, Chicago, pp. 304–320.

Lea, D.A.M. (1964). Abelam Land and Subsistence. Ph.D. Dissertation, Australian National University, Canberra.

Lee, R.B. and DeVore, I. (eds.) (1968). Man the Hunter. Aldine, Chicago.

Lee, R.B. and DeVore, I. (eds.) (1976). Kalahari Hunter-Gatherers: Studies of the !Kung San and Their Neighbors. Harvard University Press, Cambridge.

Leonzio, C., Bacci, E., Focardi, S. and Renzoni, A. (1981). Heavy metals in organisms from the northern Tyrrhenian Sea. Sci. Total Environ. 20: 131–146.

Leslie, P.W. (1985). Potential mates analysis and the study of human population structure. Yearbook Phys. Anthropol. 28: 53–78.

Lieban, R.W. (1973). Medical anthropology. In: J.J. Honigmann (eds.), Handbook of Social and Cultural Anthropology. Rand McNally, Chicago, pp. 1031–1072.

Little, M.A., Galvin, K. and Mugambi, M. (1983). Cross-sectional growth of nomadic Turkana pastoralists. Hum. Biol. 55: 811–830.

Little, M.A. and Johnson, B.R. Jr. (1986). Grip strength, muscle fatigue, and body composition in nomadic Turkana pastoralists. Am. J. Phys. Anthropol. 69: 335–344.

Littlewood, R.A. (1972). Physical Anthropology of the Eastern Highlands of New Guinea. University of Washington Press, Seattle.

Ljung, B.O., Bergstein-Brucefors, A. and Lindgren, G. (1974). The secular trend in physical growth in Sweden. Ann. Hum. Biol. 1: 245–256.

Luyken, R., Luyken-Koning, F.W.M. and Pikaar, N.A. (1964). Nutrition studies in New Guinea. Am. J. Clin. Nutr. 14: 13–27.

Maddocks, I. and Rovin, L. (1965). A New Guinea population in which blood pressure appears to fall as age advances. Papua New Guinea Med. J. 8: 17–21.

Magos, L. (1971). Selective atomic absorption determination of inorganic mercury and methylmercury in indigested biological samples. Analyst 96: 847–853.

Mahaffey, K.R., Corneliussen, P.E., Jelinek, C.F. and Fiorino, J.A. (1975). Heavy metal exposure from foods. Environ. Health Perspect. 12: 63–69.

Malcolm, L.A. (1969a). Growth and development of the Kaiapit children of the Markham valley, New Guinea. Am. J. Phys. Anthropol. 31: 39–52.

Malcolm, L.A. (1969b). Determination of the growth curve of the Kukukuku people of New Guinea from dental eruption in children and adult height. Archaeol. Phys. Anthropol. Oceania 4:72–78.

Malcolm, L.A. (1970a). Growth of the Asai child of the Madang District of New Guinea. J. Biosoc. Sci. 2: 213–226.

Malcolm, L.A. (1970b). Growth, nutrition and mortality of the infant and toddler in the Asai valley of the New Guinea Highlands. Am. J. Clin. Nutr. 23: 1090–1095.

Malcolm, L.A. (1970c). Growth and Development in New Guinea: A Study of the Bundi People of the Madang District. Monograph Series No. 1. Institute of Human Biology, Papua New Guinea, Madang.

Malcolm, L.A. (1970d). Growth and development of the Bundi child of New Guinea Highlands. Hum. Biol. 42: 293–328.

Malcolm, L.A. (1979). Protein-energy malnutrition and growth. In: F. Falkner and J.M. Tanner (eds.), Human Growth 3: Neurobiology and Nutrition. Plenum, New York, pp. 361–372.

Maleki, M. (1973). Food consumption and nutritional status of 13 year old village and city schoolboys in Fars Province, Iran. Ecol. Food Nutr. 2:

39–42.

Malina, R.M. (1975). Anthropometric correlates of strength and motor performance. *Exerc. Sport Sci. Rev.* **3**: 249–274.

Malina, R.M., Buschang, P.H., Aronson, W.D. and Selby, H.A. (1982). Aging in selected anthropometric dimensions in a rural Zapotec-speaking community in the valley of Oaxaca, Mexico. *Soc. Sci. Med.* **16**: 217–222.

Maltin, C.A., Duncan, L., Wilson, A.B. and Hesketh, J.E. (1983). Effect of zinc deficiency on muscle fiber type frequencies in the post-weanling rat. *Br. J. Nutr.* **50**: 597–604.

Mann, I. (1957). Report on ophthalamic findings in Warburton Range natives of central Australia. *Med. J. Aust.* **26**: 610-612.

Marsh, A.G., Sanchez, T.V., Michelsen, O., Chaffee, F.L. and Fagal, S.M. (1988). Vegetarian lifestyle and bone mineral density. *Am. J. Clin. Nutr.* **48**: 837–841.

Martorell, R. (1980). Interrelationship between diet, infectious disease, and nutritional status. In: L.S. Greene and F.E. Johnston (eds.), Social and Biological Predictors of Nutritional Status, Physical Growth, and Neurological Development. Academic Press, New York, pp. 81–106.

Marubini, E., Resele, L.F., Tanner, J.M. and Whitehouse, R.H. (1972). The fit of Gompertz and logistic curves to longitudinal data during adolescence on height, sitting height and biacromial diameter in boys and girls of the Harpenden growth study. *Hum. Biol.* **44**: 511–524.

Marubini, E., Resele, L.F. and Barghini, G. (1971). A comparative fitting of the Gompertz and logistic functions to longitudinal height data during adolescence in girls. *Hum. Biol.* **42**: 237–252.

Marumo, Y. (1983). Studies on trace elements in human head hair. II. Adsorption of metal elements on hair and elution from hair. *J. Hyg. Chem.* **29**: 192–198. (in Japanese with English abstract).

Masuyama, M. (1980). On the linearized growth curve. *Tokyo Rika Univ. Math.* **16**: 137–154.

Masuyama, M. (1981). Linearized growth curve for children. *Tokyo Rika Univ. Math.* **17**: 15–24.

Masuyama, M. (1982). Some analysis of the Soviet data on the height growth. *Tokyo Rika Univ. Math.* **18**: 1–22.

McAlpine, J.R. (1971). Climate of the Morehead-Kiunga area. In: Land Resources of the Morehead-Kiunga Area, Terriotory of Papua and New Guinea. Commonwealth Scientific and Industrial Research Organization (Australia), Melbourne, pp. 46–55.

McAlpine, J.R., Keig, G. and Falls, R. (1983). Climate of Papua New Guinea. Australian National Univeresity Press, Canberra.

McBean, L.D. and Speckmann, E.W. (1974). A recognition of the interrelationship of calcium with various dietary components. *Am. J. Clin. Nutr.* **27**: 603–609.

McCance, R.A. and Widdowson, E.M. (1960). The Composition of Foods. Medical Research Council Special Series No. 297. Her Majesty's Stationary Press, London.

McCarthy, F.D. and McArthur, M. (1960). The food quest and the time factor in aboriginal economic life. In: C.P. Mountford (ed.), Records of the American-Australian Scientific Expedition to Arnhem Land, Vol. 2: Anthropology and Nutrition. Melbourne University Press, Melbourne, pp.

145–194.

McKay, G.R. (1960). Growth and nutrition of infants in the Western Highlands of New Guinea. *Med. J. Aust.* **19**: 452–459.

McKenzie, J.M. (1978). Alteration of the zinc and copper concentration of hair. *Am. J. Clin. Nutr.* **32**: 470–476.

Meneely, G.R. and Battarbee, H.D. (1976). Sodium and potassium. *Nutr. Rev.* **34**: 225–235.

Meredith, H.V. (1976). Finsings from Asia, Australia, Europe, and North America on secular change in mean height of children, youths and young adults. *Am. J. Phys. Anthropol.* **44**: 315–326.

Miall, W.E., Ashcroft, M.T., Lovell, H.G. and Moore, F. (1967). A longitudinal study of the decline of adult height with age in two Welsh communities. *Hum. Biol.* **39**: 445–454.

Miall, W.E. and Chinn, S. (1973). Blood pressure and ageing: Results of a fifteen to seventeen year follow-up study in South Wales. *Clin. Sci.* **45** (Suppl. 1): 23–33.

Miall, W.E., Bell, R.A. and Lovell, H.G. (1968). Relation between change in blood pressure and weight. *Br. J. Prev. Soc. Med.* **22**: 73–80.

Minge-Klevana, W. (1980). Does labor time decrease with industrialization? A survey of time-allocation studies. *Curr. Anthropol.* **21**: 279–298.

Ministry of Education, Science and Culture of Japan (1985). School Health Statistics for 1984.

Mishra, D. and Kar, M. (1974). Nickel in plant growth and metabolism. *Bot. Rev.* **40**: 395–452.

Mittelman, D. (1980). Geometric optics and clinical refraction. In: G.A. Peyman, D.R. Sanders and M.F. Goldberg (eds.), Principles and Practice of Ophthalamology. Saunders, Philadelphia, pp. 174–221.

Moji, K. (1987). Invariable daily energy expenditure of Sundanese villagers with recent changes in their time allocation. In: T. Suzuki and R. Ohtsuka (eds.), Human Ecology of Health and Survival in Asia and the South Pacific. University of Tokyo Press, Tokyo, pp. 165–184.

Montogomery, E. (1973). Ecological aspects of health and disease in local populations. *Ann. Rev. Anthropol.* **2**: 30–35.

Moon, J., Davison, A.J., Smith, T.J. and Fadl, S. (1988). Correlation clusters in the accumulation of metals in human scalp hair: Effects of age, community of residence, and abundances of metals in air and water supplies. *Sci. Total Environ.* **72**: 87–112.

Moon, J., Smith, T.J., Tamaro, S., Enarson, D., Fadl, S., Davison, A.J. and Weldon, L. (1986). Trace metals in scalp hair of children and adults in three Alberta Indian villages. *Sci. Total Environ.* **54**: 107–125.

Morishita, H. (1965). A study on dimensional analysis of growth in Japanese: Muturation. *Res. J. Phys. Educ.* **8**: 93–99. (in Japanese).

Morishita, H. (1969). A study on relative growth of weight and height in infants. *Res. J. Phys. Educ.* **13**: 189–194. (in Japanese with English summary).

Morita, H., Shimomura, S., Kimura, A. and Morita, M. (1986). Interrelationships between the concentration of magnesium, calcium, and strontium in the hair of Japanese school children. *Sci. Total Environ.* **54**: 95–105.

Morren, G.E.J. Jr. (1974). Settlement Strategies and Hunting in a New Guinea Society. Ph.D. Dissertation, Columbia University.

Moser, P.B., Reynolds, R.D., Acharya, S., Howard, P., Andon, M.B. and

Lewis, S.A. (1988). Copper, iron, zinc, and selenium dietary intake and status of Nepalese lactating women and their breast-fed infants. *Am. J. Clin. Nutr.* **47**: 729–734.

Mulder, M.B. and Caro, T.M. (1985). The use of quantitative observational techniques in anthropology. *Curr. Anthropol.* **26**: 323–335.

Muldowney, F.P., Crooks, J. and Bluhm, M.M. (1957). The relationship of total exchangeable potassium and chloride to lean body mass, red cell mass and creatinine excretion in man. *J. Clin. Invest.* **36**: 1375–1381.

Munroe, R.H., Koel, A., Munroe, R.L., Bolton, R., Michelson, C. and Bolton, C. (1983). Time allocation in four societies. *Ethnology* **22**: 235–370.

Murai, M., Itoh, K., Ono, C., Kumagawa, K., Sakamoto, M. and Toyokawa, H. (1985). Differences between analyzed and calculated amounts of sodium in representative foods of daily meals and the availability of the sodium residual rate. *J. Jpn. Soc. Nutr. Food Sci.* **38**: 383–386. (in Japanese with English abstract).

Murai, M., Pen, F. and Miller, C.D. (1958). Some Tropical South Pacific Island Foods. University of Hawaii Press, Honolulu.

Myron, D.R., Zimmerman, T.J., Shuler, T.R., Klevay, L.M., Lee, D.E. and Nielsen, F.H. (1978). Intake of nickel and vanadium by humans: A survey of selected diets. *Am. J. Clin. Nutr.* **31**: 527–531.

Nag, M. (1962). Factors Affecting Human Fertility in Nonindustrial Societies: A Cross-Cultural Study. Yale University Publications in Anthropology No.66. Yale University Press, New Haven.

Nagahashi, M., Yamazaki, N., Ohi, G., Kai, I., Suzuki, H., Hayakawa, K., Nagasako, K. and Kimura, K. (1985). Dietary fiber intake and diverticular disease of the colon: A case control study. *Jpn. J. Hyg.* **40**: 781–788. (in Japanese with English abstract).

Nagamine, S. and Suzuki, S. (1964). Anthropometry and body composition of Japanese young men and women. *Hum. Biol.* **36**: 8–15.

National Diabetes Data Group (1979). Classification and diagnosis of diabetes mellitus and other categories of glucose intolerance. *Diabetes* **28**: 1039–1057.

National Institute of Health Consensus Development Panel on the Health Implications of Obesity (1985). Health implications of obesity: National Institute of Health Consensus Development Conference Statement. *Ann. Intl. Med.* **103**: 1073–1077.

National Research Council (1980). Recommended Dietary Allowances (9th ed.). National Academy of Sciences, Washington, D.C.

Neel, J.V. and Weiss, K. (1975). The genetic structure of a tribal population, the Yanomama Indians. XII. Biodemographic studies. *Am. J. Phys. Anthropol.* **42**: 25–52.

Neel, J.V., Layrisse, M. and Salzano, F.M. (1977). Man in the Tropics: Population Structure and Human Variation. Cambridge University Press, Cambridge, pp. 109–142.

Neel, J.V., Salzano, F.M., Junqueira, P.C., Keiter, F. and Maybury-Lewis, D. (1964). Studies on the Xavante Indians of the Brazilian Matto Grosso. *Am. J. Hum. Genet.* **16**: 52-140.

Newell, F.W. (1982). Ophthalmology: Principles and Concepts. 5th ed. Mosby, St. Louis.

Nishihara, T., Shimamoto, T., Kondo, M., Ando, Y., Kushida, I., Iio, T., Onosaka, T., Tanaka, K. and Yokoyama, T. (1979). Cadmium contents in foods and the estimation of cadmium uptake from foods. *J. Hyg. Chem.* **25**: 346–351.

Nishikawa, G., Oda, K. and Suganuma, H. (1979). Botanical description of sago palm. *Jpn. J. Trop. Agr.* **23**: 117–171. (in Japanese with English Summary).

Norgan, N.G., Ferro-Luzzi, A. and Durnin, J.V.G.A. (1974). The energy and nutrient intake and the energy expenditure of 204 New Guinea adults. *Phil. Trans. Royal Soc. Lond.* B **268**: 309–348.

Norgan, N.G., Ferro-Luzzi, A. and Durnin, J.V.G.A. (1982). The body composition of New Guinean adults in contrasting environments. *Ann. Hum. Biol.* **9**: 343–353.

Odum, E.P. (1971). Fundamentals of Ecology (3rd ed.). W.B. Saunders, Philadelphia.

Odum, H.T. (1967). Energetics of world food production. In The World Food Problem, A Report of the President's Science Advisory Comittee. Panel on World Food Supply. The White House, Washington, D.C. Vol. 3, pp. 55–94.

Ohtsuka, R. (1977a). The sago eaters: An ecological discussion with special reference to the Oriomo Papuans. In: J. Allen, J. Golson and R. Jones (eds.), Sunda and Sahul: Prehistoric Studies in Southeast Asia, Melanesia and Australia. Academic Press, London, pp. 465–492.

Ohtsuka, R. (1977b). Time-space use of the Papuans depending on sago and game. In: H. Watanabe (ed.): Human Activity System: Its Spationtemporal Structure. University of Tokyo Press, Tokyo, pp. 231–260.

Ohtsuka, R. (1983). Oriomo Papuans: Ecology of Sago-Eaters in Lowland Papua. University of Tokyo Press, Tokyo.

Ohtsuka, R. (1985). Comments on Mulder and Caro's article. *Curr. Anthropol.* **26**: 331–332.

Ohtsuka, R. (1987). Man surviving as a population: A study of the Gidra in lowland Papua. In: T. Suzuki and R. Ohtsuka (eds.), Human Ecology of Health and Survival in Asia and the South Pacific. University of Tokyo Press, Tokyo, pp. 17–34.

Ohtsuka, R. and Suzuki, T. (1978). Zinc, copper and mercury in Oriomo Papuan's hair. *Ecol. Food Nutr.* **6**: 243–249.

Ohtsuka, R. and Suzuki, T. (1989). Population ecology of human survival in diversified Melanesian environment. MAB (Researches Related to the UNESCO's Man and Biosphere Programme in Japan) 1988–1989, pp. 55–62.

Ohtsuka, R., Hongo, T., Kawabe, T., Suzuki, T., Inaoka, T., Akimichi, T. and Sasano, H. (1985). Mineral content of drinking water in tropical lowland Papua. *Env. Intl.* **11**: 505–508.

Ohtsuka, R., Suzuki, T. and Morita, M. (1986). Sodium-rich tree ash as a native salt source in lowland Papua. *Econ. Bot.* **41**: 55–59.

O'Leary, M.J., McClain, C.J. and Hegarty, P.V.J. (1979). Effect of zinc deficiency on the weight, cellularity and zinc concentration of different skeletal muscles in the postweanling rat. *Br. J. Nutr.* **42**: 487–495.

Oliveira, J.F.S., de Carvalho, J.P., de Sousa, P.F.X.B. and Simão, M.M. (1976). The nutritional value of four species of insects consumed in Angola.

Ecol. Food Nutr. **5**: 91–97.

Olson, R.E. (ed.) (1975). Protein-Calorie Malnutrition. Academic Press, New York.

Oomen, H.A.P.C. (1961). The nutrition situation in Western New Guinea. *Trop. Geogr. Med.* **13**: 321–335.

Oomen, H.A.P.C. (1971). Ecology of human nutrition in New Guinea: Evaluation of subsistence patterns. *Ecol. Food Nutr.* **1**: 1–16.

Oomen, H.A.P.C. and Grubben, G.J.H. (1978). Tropical Leaf Vegetables in Human Nutrition. Department of Agricultural Research, Amsterdam.

Page, L.B., Damon, A. and Moellering, R.C. Jr. (1974). Antecedents of cardiovascular diseases in six Solomon Islands societies. *Circulation* **49**: 1132–1146.

Pagezy, H. and Haupie, R.C. (1985). Seasonal variation in the growth rate of weight in African babies, aged 0 to 4 years. *Ecol. Food Nutr.* **18**: 29–41.

Paijmans, K., Blace, D.H. and Bleeker, P. (1971). Summary description of the Morehead-Kiunga Area, Territory of Papua and New Guinea. Commonwealth Scientific and Industrial Research Organization (Australia), Melbourne, pp. 12–18.

Pennington, J.A.T. (1987). Aluminum content of foods and diets. *Food Addit. Contam.* **5**: 161–232.

Pennington, J.A.T., Young, B.E., Wilson, D.B., Johnson, R.D. and Vanderveer, J.E. (1986). Mineral content of foods and total diets: The selected minerals in foods survey, 1982 to 1984. *J. Am. Diet. Assoc.* **86**: 876–891.

Perkons, A.K., Velandia, J.A. and Dienes, M. (1977). Forensic aspects of trace element variation in the hair of isolated Amazonas Indian Tribes. *J. Forensic Sci.* **22**: 95–105.

Peters, F.E. (1957). Chemical composition of South Pacific foods: An annotated bibliography. Technical Paper 100. South Pacific Commission, Noumea.

Phillips, D.J.H. (1977). The use of biological indicator organisms to monitor trace metal pollution in marine and estuarine environments: A review. *Environ. Pollut.* **13**: 281–317.

Piotrowski, J.K. and Inskip, M.J. (1981). Health Effects of Methylmercury. MARC Report No. 24. Monitoring and Assessment Research Centre. Chelsea College, University of London.

Post, R.H. (1962). Population differences in vision acuity: A review, with speculative notes on selection relaxation. *Eugen. Quat.* **9**: 189-212.

Powdermaker, H. (1931). Vital statistics of New Ireland (Bismarck Archipelago) as revealed in genealogies. *Hum. Biol.* **3**: 351–375.

Powell. J.M. (1976). Ethnobotany. In: K. Paijmans (ed.), New Guinea Vegetation, Australian National University Press, Canberra, pp. 106–183.

Prader, A., Tanner, J.M. and von Harnack, G.A. (1963). Catch-up growth following illness or starvation. *J. Paediatr.* **62**: 646–659.

Preece, M.A. and Baines, M.J. (1978). A new family of mathematical models describing the human growth curve. *Ann. Hum. Biol.* **5**: 1–24.

Pressat, R. (1972). Demographic Analysis: Methods, Results, Applications. Aldine-Atherton, Chicago.

Pressat, R. (1985). Completed fertility. In: C. Wilson (ed.), The Dictionary of Demography. Basil Blackwell, Oxford, p. 37.

Prior, I. A. M., Evans, J. G., Harvey, H. P. B., Davidson, F. and Lindsey, U. M. (1968). Sodium intake and blood pressure in two Polynesian populations. *New Engl. J. Med.* **279**: 515–520.

Purseglove, J.W. (1972). Tropical Crops: Monocotyledons. Longman, London.

Rappaport, R.A. (1968). Pigs for the Ancestors: Ritual in the Ecology of a New Guinea People. Yale University Press, New Haven.

Rarick, G.L., Wainer, H., Thissen, D. and Seefeldt, V. (1975). A double logistic comparison of growth patterns of normal children with Down's syndrome. *Ann. Hum. Biol.* **2**: 339–346.

Reeve, E.C.R. and Huxley, J.S. (1945). Some problems in the study of allometric growth. In: W.E. Le Gros Clark and P.B. Medawar (eds.), Essays on Growth and Form Presented to D'Arcy Wentworth Thompson. Clarendon, Oxford, pp. 121–156.

Reh, E. (1962). Manual on Household Food Consumption Survey. Food and Agriculture Organization of the United Nations, Rome.

Reinhold, J.G. (1971). High phytate content of rural Iranian bread: a possible cause of human zinc deficiency. *Am. J. Clin. Nutr.* **24**: 1204–1206.

Reinhold, J.G. (1972). Phytate concentrations of leavened and unleavened Iranian breads. *Ecol. Food Nutr.* **1**: 187–192.

Reinsh, C.H. (1967). Smoothing by spline functions. *Numer. Math.* **10**: 177–183.

Resources Council (1982). Standard Table of Food Composition in Japan. Fourth Revised Edition. Science and Technology Agency, Tokyo.

Richards, A.I. (1939). Land, Labour and Diet in Northern Rhodesia (2nd ed.). Oxford University Press, London.

Rivlin, R.S. (1983). Misuse of hair analysis for nutritional assessment. *Am. J. Med.* **75**: 489–493.

Ross, A.C. (1984). Migrants from Fifty Villages. Monograph 21. Institute of Applied Social and Economic Research, Boroko (Papua New Guinea).

Ross, J., Gibson, R.S. and Sabry, J.H. (1986). A study of seasonal trace element intakes and hair trace element concentrations in selected households from the Wosera, Papua New Guinea. *Trop. Geogr. Med.* **38**: 246–254.

Sagami, Y. (1967). Relative growth of human mandible. *Acta Anat. Nippon* **42**: 240–257. (in Japanese with English summary).

SAS Institute (1981). SAS/GRAPH User's Guide, 1981 edition. SAS Institute, North Carolina.

SAS Institute (1982). SAS User's Guide: Basics, Statistics, 1982 edition. SAS Institute, North Carolina.

Sasaki, N., Takemori, K., Ohtsuka, R. and Suzuki, T. (1981). Mineral contents in hair from Oriomo Papuans and Akita dwellers. *Ecol. Food Nutr.* **11**: 117–120.

Sato, K. (1947). A study on allometry in Japanese boys. *Igaku To Seibutsugaku* **11**: 329–330. (in Japanese).

Schelenz, R. (1977). Dietary intake of 25 elements by man estimated by neuron activation analysis. *J. Radioanal. Chem.* **37**: 539–548.

Schreider, E. (1964). Ecological rules, body heat regulation and human evolution. *Evolution* **18**: 1–19.

Shephard, R.J. (1985). Some limitation of exercise testing. *J. Sport Med. Phys. Fitness* **25**: 40–48.

Shepherd, R.H., Sholl, D.A. and Vizozo, A. (1949). The size relationships subsisting between body length, limbs and jaws in man. *J. Anat.* **83**:

296–302.

Sherlock, J.C., Lindsay, D.G., Hislop, J.E., Evans, W.E. and Collier, T.R. (1982). Duplication diet study on mercury intake by fish consumers in the United Kingdom. *Arch. Environ. Health* **37**: 271–278.

Shimizu, M. (1959). Relative Growth. Kyodoisho Shuppan, Tokyo. (in Japanese).

Shimizu, M. and Inoue, S. (1956). Relative growth of Japanese school children. *Shinshu Med. J.* **5**: 251–256. (in Japanese).

Shirai, F. (1988). Studies on the relationship between the characteristics of daily food consumption and the concentration of mercury in urine. *Jpn. J. Hyg.* **43**: 923–933. (in Japanese with English abstract).

Shiraishi, K., Kawamura, H., Ouchi, M. and Tanaka, G. (1986). Determination of 15 mineral elements in total diet samples by inductively coupled plasma atomic emission spectroscopy. *J. Jpn. Soc. Nutr. Food Sci.* **39**: 209–215. (in Japanese with English abstract).

Shishido, S. and Suzuki, T. (1974). Estimation of daily intake of inorganic or organic mercury via diet. *Tohoku J. Exp. Med.* **114**: 369–377.

Simmons, W.K. (1972). Urinary urea nitrogen/creatinine ratio as indicator of recent protein intake in field studies. *Am. J. Clin. Nutr.* **25**: 539–542.

Sinnett, P.F. (1975). The People of Murapin. Monograph Series No. 4. Papua New Guinea Institute of Medical Research, Goroka (Papua New Guinea).

Sinnett, P.F. and Whyte, H.M. (1973a). Epidemiological studies in a Highland population of New Guinea: Environment, culture, and health status. *Hum. Ecol.* **1**: 245–277.

Sinnett, P.F. and Whyte, H.M. (1973b). Epidemilogical studies in a total highland population, Tukisenta, New Guinea: Cardiovascular disease and relevant clinical electrocardiographic, radiological and biochemical findings. *J. Chron. Dis.* **26**: 265–290.

Siwani, I.N. (1981). Personal communication.

Skeldon, R. (1979). Internal migration. In: R. Skeldon (ed.), The Demography of Papua New Guinea: Analyses from the 1971 Census. Monograph 11. Institute of Applied Social and Economic Research, Boroko (Papua New Guinea), pp. 77–110.

Skolnick, M., Cavalli-Sforza, L.L., Moroni, A. and Siri, E. (1976). A preliminary analysis of the genealogy of Parma Valley, Italy. In: R.H. Ward and K.M. Weiss (eds.), The Demographic Evolution of Human Populations. Academic Press, London, pp. 95–115.

Smith, E.A. (1979). Human adaptaion and energetics. *Hum. Ecol.* **7**: 53–74.

Snedecor, G.W. and Cochran, W.G. (1980). Statistical Methods (7th ed.). Iowa State University Press, Iowa.

Solomons, N.W. (1982). Biological availability of zinc in humans. *Am. J. Clin. Nutr.* **35**: 1048–1075.

Solomons, N.W. and Jacob, R.A. (1981). Studies on the bioavailability of zinc in humans: Effects of heme and nonheme iron on the absorption of zinc. *Am. J. Clin. Nutr.* **34**: 475–482.

Sorentino, C. (1979). Mercury in marine and freshwater fish of Papua New Guinea. *Aust. J. Mar. Freshwater Res.* **30**: 617–623.

Spring, J.A., Robertson, J. and Buss, D.H. (1979). Trace nutrients. 3: Magnesium, copper, zinc, vitamin B_6, vitamin B_{12} and folic acid in the British household food supply. *Br. J. Nutr.* **41**: 487–493.

SPSS-X™ User's Guide (1988). 3rd ed. SPSS, Chicago, pp. 480–498.

Stauber, J.L. Florence, T.M. and Webster, W.S. (1987). The use of scalp hair to monitor manganese in Aborigines from Groote Eylandt. *Neurotoxicology* **8**: 431–435.

Stephens, M.A. (1974). EDF statistics for goodness of fit and some comparisons. *J. Am. Stat. Assoc.* **69**: 730–737.

Strain, W.H., Steadman, L.T., Lankau, C.A. Jr., Berliner, W.P. and Pories, W.J. (1966). Analysis of zinc levels in hair for the diagnosis of zinc deficiency in man. *J. Lab. Clin. Med.* **68**: 244–249.

Suzuki, T. (1977). Human ecology in understanding environmental health problem. *Rev. Environ. Health* **2**: 204–219.

Suzuki, T. (1980). Methodology of Human Ecology. University of Tokyo Press, Tokyo. (in Japanese).

Suzuki, T. (1981). How great will the stature of Japanese eventually become? *J. Hum. Ergol.* **10**: 13–24.

Suzuki, T. (1984). What is nutritional ecology? In: H. Koishi and T. Suzuki (eds.), Nutritional Ecology. Kowa-shuppan, Tokyo, pp. 9–24. (in Japanese).

Suzuki, T. (1985). The traditional art of curing among the Gidra of the Oriomo Plateau, Papua New Guinea. *Man Cul. Oceania* **1**: 67–79.

Suzuki, T. (1987). Population and ecosystem in the study of health and survival strategy. In: T. Suzuki and R. Ohtsuka (eds.), Human Ecology of Health and Survival in Asia and the South Pacific. University of Tokyo Press, Tokyo, pp. 3–14.

Suzuki, T. (1988). Hair and nails: Advantages and pitfalls when used in biological monitoring. In: T.W. Clarkson, L. Friberg, G.F. Nordberg and P.R. Sager (eds.), Biological Monitoring of Toxic Metals. Plenum, New York, pp. 623–640.

Suzuki, S. and Lu, C.C. (1976). A balance study of cadmium: An estimation of daily input, output and retained amount in two subjects. *Ind. Health* **14**: 53–65.

Suzuki, T., Akimichi, T., Kawabe, T., Inaoka, T. and Ohtsuka, R. (1984a). Growth of the Gidra in lowland Papua New Guinea. In: N. Kobayashi and T.B. Brazelton (eds.), The Growing Child in Family and Society. University of Tokyo Press, Tokyo, pp. 77–93.

Suzuki, T., Hongo, T., Morita, M. and Yamamoto, R. (1984b). Elemental contamination of Japanese women's hair from historical samples. *Sci. Total Environ.* **39**: 81–91.

Swadling, P. (1983). How long people have been in the Ok Tedi Impact Region? Papua New Guinea National Museum Record No. 8, Port Moresby.

Swift, C.E. and Berman, M.D. (1959). Factors affecting the water retention of beef. I: Variations in composition and properties among eight muscles. *Food Technol.* **13**: 365–370.

Taber, L.A.L. and Cook, R.A. (1980). Dietary and anthropometric assessment of adult omnivores, fish-eaters, and lacto-ovo-vegetarians. *J. Am. Diet. Assoc.* **76**: 21–29.

Takagi, Y., Matsuda, S., Imai, S., Ohmori, Y., Masuda, T., Vinson, J.A., Mehra, M.C., Puri, B.K. and Kaniewski, A. (1986). Trace elements in human hair: An international comparison. *Bull. Environ. Contam. Toxicol.* **36**: 793–800.

Takemori, K. (1976). Studies on sodium content in human hair. Part 3: The mechanisms of the external contamination of hair samples. *Hirosaki Med. J.* **28**: 68–76. (in Japanese with English abstract).

Takemori, K. (1980). Collection and transport of urine samples by filter paper absorption method: The methods of the determination of sodium, potassium and creatinine concentration in urine. *Jpn. J. Hyg.* **11**: 721–727. (in Japanese with English summary).

Takeuchi, T., Hayashi, T., Takada, J., Hayashi, Y., Koyama, M., Kozuka, H., Tsuji, H., Kusaka, Y., Ohmori, S., Shinogi, M., Aoki, A., Katayama, K. and Tomiyama, T. (1982). Variation of elemental concentration in hair of the Japanese in terms of age, sex and hair treatment. *J. Radioanal. Chem.* **70**: 29–55.

Tanaka, J. (1980). The Sun: Hunter-Gatherers of the Kalahari. Univesity of Tokyo Press, Tokyo.

Tanaka, R., Ikebe, K., Tanaka, Y. and Kunita, N. (1983). Heavy metals contents in processed foods and daily intake of heavy metals by foodstuffs in Japanese. *J. Food Hyg. Soc. Jpn.* **24**: 488–499. (in Japanese).

Tanner, J.M. (1947). The morphological level of personality. *Proc. Roy. Soc. Med.* **40**: 301-308.

Tanner, J.M. (1951). Some notes on the reporting of growth data. *Hum. Biol.* **23**: 93–159.

Tanner, J.M. (1962). Growth at Adolescence (2nd ed.). Blackwell Scientific Publications, Oxford.

Tanner, J.M. (1968). Earlier maturation in man. *Sci. Am.* **218**: 21–27.

Tanner, J.M. (1977). Human growth and constitution. In: G.A. Harrison, J.S. Weiner, J.M. Tanner and N.A. Barnicot, Human Biology: An Introduction to Human Evolution, Variation, Growth and Ecology. Oxford University Press, Oxford, pp. 301–385.

Tanner, J.M. (1978). Foetus into Man: Physical Growth from Conception to Maturity. Harvard University Press, Cambridge.

Tanner, J.M. and Cameron, N. (1980). Investigation of the midgrowth spurt in height, weight and limb circumferences in single-year velocity data from the London 1966-67 Growth Survey. *Ann. Hum. Biol.* **7**: 565–577.

Tanner, J.M., Healy, M.J.R., Lockhart, R.D., Mackenzie, J.D. and White-house, R.H. (1956). Aberdeen Growth Study. I: The prediction of adult body measurements from measurements taken each year from birth to 5 years. *Arch. Dis. Childhood* **31**: 372–381.

Tanner, J.M., Whitehouse, R.H., Marubini, E. and Resele, L.F. (1976). The adolescent growth spurt of boys and girls of the Harpenden Growth Study. *Ann. Hum. Biol.* **3**: 109–126

Tanner, J.M., Whitehouse, R.H. and Takaishi, M. (1966). Standards from birth to maturity for height, weight, height velocity, and weight velocity: British children, 1965, Part I, Part II. *Arch. Dis. Childhood* **41**: 454–471, 613–635.

Teramoto, K., Horiguchi, S. and Ninomiya, K. (1987). Zinc content in Japanese food and estimated average daily intake. *Asia-Pacific J. Public Health* **1**: 32–42.

Teraoka, H., Morii, F. and Kobayashi, J. (1981). The concentrations of 24 elements in foodstuffs and the estimate of their daily intake. *J. Jpn. Soc. Nutr. Food Sci.* **34**: 221–239. (in Japanese with English abstract).

Thissen, D., Bock, R.D., Wainer, H. and Roche, A.F. (1976). Individual growth in stature: A comparison of four growth studies in the U.S.A. *Ann. Hum. Biol.* **3**: 529–542.

Thomas, R.B. (1973). Human Adaptation to a High Andean Energy-Flow System. Occasional Papar in Anthropology 7. Pennsilvania State University Department of Anthropology, Univerisity Park.

Thompson, D'Arcy W. (1942). On Growth and Form. Cambridge University Press, Cambridge.

Thompson, P., Rosenborough, R., Russek, E., Jacobson, M. and Moser, P.B. (1986). Zinc status and sexual development in adolescent girls. *J. Am. Diet. Assoc.* **86**: 892–897.

Thomson, A.B.R., Olatunbosun, D. and Valberg, L.S. (1971). Interrelation of intestinal transport system for manganese and iron. *J. Lab. Clin. Med.* **78**: 642–655.

Tobian, L. (1979). The relationship of salt to hypertension. *Am. J. Clin. Nutr.* **32**: 2739–2748.

Townsend, P.K. (1971). New Guinea sago gatherers: A study of demography in relation to subsistence. *Ecol. Food Nutr.* **1**: 19–24.

Townsend, P.K. (1974). Sago production in a New Guinea economy. *Hum. Ecol.* **2**: 217–236.

Truswell, A.S., Kennerlly, B.M., Hansen, J.D.L. and Lee, R.B. (1972). Blood pressure of !Kung Bushmen in northern Botsuwana. *Am. Heart J.* **84**: 5–12.

Tsugane, S. (1986). The content of chemical elements in the hair of Japanese immigrants in four different places in South America. *Jpn. J. Public Health* **33**: 535–546. (in Japanese with English abstract).

Turnlund, J.R., King, J.C., Keyes, W.R., Gong, B. and Michel, M.C. (1984). A stable isotope study of zinc absorption in young men: Effects of phytate and α-cellulose. *Am. J. Clin. Nutr.* **40**: 1071–1077.

Tylavsky, F.A. and Anderson, J.J.B. (1988). Dietary factors in bone health of elderly lactoovovegetarian and omnivorous women. *Am. J. Clin. Nutr.* **48**: 842–849.

Tyroler, H.A., Heyden, S. and Hames, C.G. (1975). Weight and hypertension. Evans County studies of Blacks and Whites. In: O. Paul (ed.), Epidemiology and Control of Hypertension. Simposia Specialists, Miami, pp. 177–201.

Ulijaszek, S.J. (1982). Nutritional status of a sago-eating community in the Purari Delta, Gulf Province. IASER Discussion Paper No. 44, Institute of Applied Social and Economic Research, Boroko (Papua New Guinea).

United Nations (1982). Model Life Tables for Developing Coutries. United Nations, New York.

Vanderkooy, P.D.S. and Gibson, R.S. (1987). Food consumption patterns of Canadian preschool children in relation to zinc and growth status. *Am. J. Clin. Nutr.* **45**: 609–616.

Van Wieringen, J.C. (1978). Secular growth changes. In: F. Falkner and J.M. Tanner (eds.), Human Growth. 2: Postnatal Growth. Plenum, New York, pp. 445–473.

Varo, P. and Koivistoinen, P. (1980). Mineral element composition of Finnish foods. XII: General discussion and nutritional evaluation. *Acta Agr. Scand.* Suppl. **22**: 165–171.

Venkatachalam, P.S. (1962). A study of the diet, nutrition and health of the people of the Chimbu area (New Guinea Highlands). Department of Public

Health Monograph No. 4, Territory of Papua and New Guinea, Port Moresby.

Vines, A.P. (1970). An Epidemiological Sample Survey of the Highlands, Mainland and Islands Regions of the Territory of Papua and New Guinea. Department of Public Health, Papua and New Guinea, Port Moresby.

Walsh, R.J., Murrell, T.G.C. and Bradley, M.A. (1966). A medical and blood group survey of the Lake Kopiago natives. *Archaeol. Phys. Anthropol. Oceania* **1**: 57–66.

Wark, L. and Malcolm, L.A. (1969). Growth and development of the Lumi child in the Sepik District of New Guinea. *Med. J. Aust.* **2**: 129–136.

Watanabe, H. (1971). Periglacial ecology and the emergence of *Homo sapiens*. In: The Origin of *Homo sapiens* (Ecology and Conservation). UNESCO, Paris, pp. 271–285.

Watanabe, H. (1975). Bow and Arrow Census in a West Papuan Lowland Community: A New Field for Functional-Ecological Study. Occasional Paper in Anthropology No. 5. University of Queensland, St. Lucia.

Watanabe, H. (ed.) (1977). Human Activity System: Its Spatiotemporal Structure. University of Tokyo Press, Tokyo.

Weiner, J.S. (1977). Nutritional ecology. In: G.A. Harrison, J.S. Weiner, J.M. Tanner and N.A. Barnicot, Human Biology (2nd ed.). Oxford University Press, Oxford, pp. 400–423.

Weiss, K.M. (1972). A general measure of human population growth regulation. *Am. J. Phys. Anthropol.* **37**: 337–344.

Wellin, E. (1978). Theoretical orientation in medical anthropology: Change and continuity over the past half century. In: M.H. Logan amd E.E. Hunt, Jr. (eds.), Health and the Human Condition. Duxbury Press, North Scituate (Massachusetts), pp. 23–39.

Wenlock, R.W., Buss, D.H. and Dixon, E.J. (1979). Trace nutrients. 2: Manganese in British food. *Br. J. Nutr.* **41**: 253–261.

WHO (1972). Evaluation of certain food additives and the contaminants, mercury, lead and cadmium. Sixteenth Report of the Joint FAO/WHO Expert Committee on Food Additives. WHO Technical Report, Series No. 505. World Health Organization, Geneva.

WHO (1980). Consultation to Re-Examine the WHO Environmental Health Criteria for Mercury. World Health Organization, Geneva.

Wilkinson, S.R., Stuedemann, J.A., Grunes, D.L. and Devine, O.J. (1987). Relation of soil and plant magnesium to nutrition of animal and man. *Magnesium* **6**: 74–90.

Williams, F.E. (1936). Papuans of the Trans-Fly. Clarendon, Oxford.

Wilson, C.S. (1973). Food taboos of childbirth: The Malay example. *Ecol. Food Nutr.* **2**: 267–274.

Winterhalder. B. and Smith E.A. (eds.) (1981). Hunter-Gatherer Foraging Strategies: Ethnographic and Archeological Analyses. University of Chicago Press, Chicago.

Wobst, H.M. (1974). Boundary conditions for Paleolithic social systems: A simulation approach. *Am. Antiquity* **39**: 147–178.

Wolstenholme, J. and Walsh, R.J. (1967). Heights and weights of indigens of the Western Highlands District, New Guinea. *Archaeol. Phys. Anthropol. Oceania* **2**: 220-226.

Working Group on Mercury in Fish (1980). Report on Mercury in Fish and

Fish Products. Australian Government Publishing Service, Canberra.

Workman, P.L. and Devor, E.J. (1980). Population and society in Aland, Finland 1760-1880. In: B. Dyke, and W.T. Morill (eds.), Genealogical Demography. Academic Press, New York, pp. 179–196.

Wurm, S.A. (1971). Notes on the linguistic situation in the Trans-Fly area. *Papers in New Guinea Linguistics* **14**: 115–172.

Wurm, S.A. (1982). Papuan Languages of Oceania. Gunter Narr Verlag Tübingen, Stuttgart.

Xue-Cun, C., Tai-An, Y., Jin-Sheng, H., Qui-Yan, M., Zhi-Min, H. and Li-Xiang, L. (1985). Low levels of zinc in hair and blood, pica, anorexia and poor growth in Chinese preschool children. *Am. J. Clin. Nutr.* **42**: 694–700.

Yamagata, N. (1977). Trace Elements. Sangyo Tosho, Tokyo. (in Japanese).

Yamamoto, R., Satoh, H., Suzuki, T., Naganuma, A. and Imura, N. (1980). The applicable condition of Magos' method for mercury measurement under coexistence of selenium. *Anal. Biochem.* **101**: 254–259.

Yamori, Y., Kihara, M., Fujikawa, J., Soh, Y., Nara, Y., Ohtaka, M., Horie, R., Tsunematsu, T., Note, S. and Fukase, M. (1982a). Dietary risk factors of stroke and hypertension in Japan. Part 1: Methodological assessment of urinalysis for dietary salt and protein intakes. *Jpn. Cir. J.* **46**: 933–938.

Yamori, Y., Kihara, M., Fujikawa, J., Soh, Y., Nara, Y., Ohkata, M., Horie, R., Tsunematsu, T., Note, S. and Fukase, M. (1982b). Dietary risk factors of stroke and hypertension in Japan. Part 2: Validity for dietary salt and protein intakes under a field condition. *Jpn. Cir. J.* **46**: 939–943.

Yen, D.E. (1974). Arboriculture in the subsistence of Santa Cruz, Solomon Islands. *Econ. Bot.* **28**: 247–284.

Yomota, C. Isshiki, K., Kato, T., Kamikura, M., Shiroishi, Y., Nishijima, M., Hayashi, H., Husakawa, Y., Yokoyama, T., Yoneda, M., Moriguchi, H., Uchiyama, H., Shiro, T. and Ito, Y. (1987). Estimation of daily intake of vitamins, minerals and free amino acids from fresh foods purchased in Japan according to the market basket method. *J. Jpn. Soc. Nutr. Food Sci.* **40**: 451–456. (in Japanese with English abstract).

Yoshida, S. and Ikebe, K. (1988). Correlation between analyzed and calculated values of daily sodium and potassium intakes by duplicate portion study during a one-week period. *J. Jpn. Soc. Nutr. Food Sci.* **41**: 315–319. (in Japanese with English abstract).

Yost, J.A. and Kelley, P.A. (1983). Shotguns, blowguns, and spears: The analysis of technological efficiency. In: R.B. Hames and W.T. Vickers (eds.), Adaptive Responses of Native Amazonians. Academic Press, New York, pp. 189–224.

SOURCE MATERIALS BY CHAPTER
(Numbers refer to chapter in this volume.)

2. Kawabe, T., Ohtsuka, R., Inaoka, T., Akimichi, T., and Suzuki, T. (1985). Visual acuity of the Gidra in lowland Papua New Guinea. *Journal of Biosocial Science,* **17:** 361–369.
3. Ohtsuka, R., Inaoka, T., Kawabe, T., and Suzuki, T. (1987). Grip strength and body composition of the Gidra Papuans in relation to ecological conditions. *Journal of Anthropological Society of Nippon,* **95:** 457–467.
4. Ohtsuka, R. (1989). Hunting activity and aging among the Gidra Papuans: A biobehavioral analysis. *American Journal of Physical Anthropology,* **80:** 31–39.
5. Ohtsuka, R. (1985) The Oriomo Papuans: Gathering versus horticulture in an ecological context. In: V.N. Misra and P. Bellwood (eds.), *Recent Advances in Indo-Pacific Prehistory.* Oxford & IBH, New Delhi, pp. 343–348.
7. Ohtsuka, R., Kawabe, T., Inaoka, T., Suzuki, T., Hongo, T., Akimichi, T., and Sugahara, T. (1984). Composition of local and purchased foods consumed by the Gidra in lowland Papua. *Ecology of Food and Nutrition,* **15:** 159–169.*
8. Hongo, T., Suzuki, T., Ohtsuka, R., Kawabe, T., Inaoka, T., and Akimichi, T. (1989). Compositional character of Papuan foods. *Ecology of Food and Nutrition,* **23:** 39–56.*
9. Ohtsuka, R., Inaoka, T., Kawabe, T., Suzuki, T., Hongo, T., and Akimichi, T. (1985). Diversity and change of food consumption and nutrient intake among the Gidra in lowland Papua. *Ecology of Food and Nutrition,* **16:** 339–350.*
10. Hongo, T., Suzuki, T., Ohtsuka, R., Kawabe, T., Inaoka, T., and Akimichi, T. (1989). Element intake of the Gidra in lowland Papua: Inter-village variation and the comparison with contemporary levels in developed countries. *Ecology of Food and Nutrition,* **23:** 293–309.*
11. Kawabe, T. (1986). Intrapopulation variation of body physique in the Gidra, Papua New Guinea. *Man and Culture in Oceania,* **2:** 27–55.
12. Akimichi, T. (1987). Individual variation and short-term fluctuation in child growth among the Gidra in lowland Papua New Guinea. *Man and Culture in Oceania,* **3:** 125–134.
14. Inaoka, T., Suzuki, T., Ohtsuka, R., Kawabe, T., Akimichi, T., Takemori, K., and Sasaki, N. (1987). Salt consumption, body fatness and blood pressure of the Gidra in lowland Papua. *Ecology of Food and Nutrition,* **20:** 55–66.*
17. Suzuki, T., Watanabe, S., Hongo, T., Kawabe, T., Inaoka, T., Ohtsuka,

R., and Akimichi, T. (1988). Mercury in scalp hair of Papuans in the Fly estuary, Papua New Guinea. *Asia-Pacific Journal of Public Health*, **2**: 39–47.

18. Ohtsuka, R. (1986). Low rate of population increase of the Gidra Papuans in the past: A genealogical-demographic analysis. *American Journal of Physical Anthropology*, **71**: 13–23.

20. Ohtsuka, R., Kawabe, T., Inaoka, T., Akimichi, T., and Suzuki, T. (1985). Inter- and intra-population migration of the Gidra in lowland Papua: A pupulation-ecological analysis. *Human Biology*, **57**: 33–45.

SUBJECT INDEX

261

POPULATION INDEX